The book series, Human-Environment Interactions, is designed to publish the best research being produced on the reciprocal interactions between human populations and the biophysical environment. This scholarship is inherently multidisciplinary, combining contributions from the physical, biological, and social sciences, coalescing around investigations into the human dimensions of global environmental change. Because this is an integrated science agenda, the books in the series will reflect this multidisciplinary complexity.

The series publishes work that addresses issues both timely and urgent in nature. They speak to our future, our present, and our past. In order to better address the ways in which feedbacks operate between the biophysical environment and the social and political context of human choice-making, series books provide a nuanced understanding, solidly backed by quantitative science, of the interactions of population and environment, their changing spatial and temporal distribution, their consumption patterns and their impact, and their changing production and reproduction strategies. Other issues that concern the series are the health impacts of global warming and the distribution of disease vectors; the impact of El Niño events upon coastal areas and upon places as varied as the desert Southwest and the Amazon; and how people forest and reforest their landscape. How might we anticipate these changes? How might we respond and prepare for these changes? Who is most vulnerable? These and other related topics will drive the series as we seek to stimulate readers to think, and to act.

Cattle Bring Us to Our Enemies

Turkana Ecology, Politics, and Raiding in a Disequilibrium System

J. Terrence McCabe

THE UNIVERSITY OF MICHIGAN PRESS

Ann Arbor

TO JIM ELLIS, IN MEMORIAM,

*intellectual leader in the
development of theory on nonequilibrium ecosystems,
mentor, colleague, and friend.*

Published in the United States of America by
The University of Michigan Press
Manufactured in the United States of America
♾ Printed on acid-free paper
2007   2006   2005   2004   4   3   2   1

A CIP catalog record for this book is available from the British Library.

Library of Congress Cataloging-in-Publication Data

McCabe, J. Terrence.
Cattle bring us to our enemies : Turkana ecology,
politics, and raiding in a disequilibrium system / J. Terrence McCabe.
p.   cm. — (Human-environment interactions)
Includes bibliographical references and index.
ISBN 0-472-09878-0 (cloth : alk. paper) —
ISBN 0-472-06878-4 (pbk. : alk. paper)
1. Turkana (African people)—Domestic animals.   2. Turkana (African
people)—Social conditions.   3. Turkana (African people)—Land tenure.
4. Cattle herding—Kenya—Turkana District.   5. Cattle stealing—Kenya—
Turkana District.   6. Turkana District (Kenya)—Social life and customs.
7. Turkana District (Kenya)—Environmental conditions. I. Title. II. Series.

DT433.545.T87M33   2004
333.74'137'0967627—dc22   2004008130

Series Editor's Introduction

The Human-Environment Interactions series is designed to publish the best scholarship on the reciprocal interactions between human populations and the biophysical environment. Inherently multidisciplinary, combining the physical, biological, and social sciences, this scholarship focuses on "the human dimensions of global environmental change." A measure of its importance is the existence over the past decade of a standing committee at the National Research Council/National Academy of Sciences that has identified and provided direction to research on the human dimensions of global change. The U.S. Global Change Research Program has been funded at significant levels by the National Science Foundation (NSF), the National Institutes of Health (NIH), the National Aeronautics and Space Administration (NASA), the National Oceanic and Atmospheric Administration (NOAA), the Department of Energy (DOE), the Environmental Protection Agency (EPA), and many other U.S. agencies. This effort is also international, with notable funding from the European Union and many other nations, such as Japan, Russia, and Brazil.

As the importance of human interactions with the earth system began to be recognized, a large community of scholars undertook a vigorous research agenda, combining methods from the biophysical and the social sciences to understand these interactions. Human-environment interactions research has addressed such concerns as tropical deforestation; the societal impacts of climate change; the reciprocal interactions of population-environment-consumption; and mega-urbanization dynamics; it has also provided large-scale monitoring of changes in vegetation and historical reconstructions of human interactions at local and regional scales. Topics such as these give scope and structure to the series. This is an integrated science agenda developed in multidisciplinary fashion because of the complex nature of the problems being tackled, and the books in the series will reflect this multidisciplinary complexity. Over the coming decade, this research will only grow.

The book that is before you reflects just such research. The South Turkana Ecosystem Project is one of the earliest and best exemplars of this marriage of first-rate ecosystem ecology research and sophisticated social science methods. Over the past decade, scientists working on the Turkana project have published a very large and comprehensive set of articles that address grassland ecosystem ecology, plant-animal interactions, and human dimensions (social organization, settlement patterns, nutrition, and politics). A particularly important contribution of the Turkana Ecosystem Project is the painstaking documentation of the behavior of these systems over time. While in the past these systems were viewed as systems in equilibrium, the scientists have shown that grassland ecology in these semiarid lands of East Africa is characterized, rather, by disequilibrium conditions. This has changed the way we think about pastoral behavior and about its role in sustaining these wildly fluctuating systems.

This book makes important contributions to several areas. It addresses the human dimensions of global change and ecological anthropology through its careful documentation over very long periods of time of the migration patterns of the Turkana and the Turkana's herd management and land use. The author examines climate factors and counterintuitive findings about reducing environmental risk by saving the most productive areas for last, rather than accessing them in the early stages of herd movement. The role of political factors, particularly violence and raiding, in pastoral decision making is studied, and it is shown that the path chosen for migration is not always the optimal one environmentally but instead is a path selected to reduce risk and to avoid predictable violence. In its investigation of violence and politics, subjects overlooked by many scholars, this book stands out as a major contribution.

Also somewhat unique as a contribution to pastoral ecology is the opportunity the author gives us to get to know several individuals and to appreciate the difficulties and challenges these persons face each day. This is fairly rare in ecology and environmental anthropology, which have in the past tended to provide aggregate statistics rather than careful life histories of individuals struggling to make life and death decisions in an uncertain environment. We see the numerous methods the Turkana use to cope with their complex social and physical situations. Each story is unique, yet from them the author is able to distill broad theoretical principles. The migration data also are unusually detailed. Although the sample is small, it provides a much richer understanding of the movements of herds and people than would have been the case with an aggregate account.

I am delighted to introduce this fine book to our readers. Books such as this are valuable because they raise issues that are timely and urgent in nature, speaking to our future and also to our past. Pastoral people have often been misunderstood and fenced out of their needed grasslands. We learn how sophisticated their management is and how much countries with pastoral peoples would gain from allowing them to move as needed. Through its examination of the interactions of population and environment, this book provides a nuanced understanding, solidly backed by quantitative science, of the feedback between the biophysical environment and human choice making in social and political contexts.

It is my hope that this book will inspire readers to commit themselves to better understanding and to acting to ensure the sustainability of ecosystems such as this one. We invite readers to send to the editor, or to members of the editorial board, proposals for books that seek to advance our knowledge of human-environment interactions.

Emilio F. Moran, Series Editor

Contents

PART 4

◈

Illustrations

Maps

Tables

Preface and Acknowledgments

I begin this book by recalling a memory that will be forever etched in my mind. It is March 1980, and I am making my first journey into Turkana District in northwest Kenya. I had read all of Philip Gulliver's work on the Turkana and was struck by how much difficulty he had in working with this group of pastoral people. In addition, while I was in Nairobi waiting for research permission and purchasing equipment, a story appeared in the *Daily Nation* about a group making a circumnavigation of Lake Turkana. When they were in southern Turkana, the place where I was to conduct my research, the headline read: "The Suguta Valley—A land where even the flies have fled." The article described the extremely harsh climate and the dangers posed by intertribal raiding and bandits. To say I was apprehensive is certainly an understatement.

I encountered no Turkana people as I was driving down the sand road between Lokichar and Lokori that day, and I stopped and looked out over the sand and gravel plains to the distant mountains. I remember thinking, Somewhere out there are Turkana families. Somehow I have to meet them and convince them that I should be allowed to live with them and to study their livestock management practices. This was how I began a period of sixteen years of research as part of the South Turkana Ecosystem Project.

Being part of the South Turkana Ecosystem Project allowed me to work closely with some of the most respected ecologists and anthropologists specializing in the study of pastoralist peoples and arid lands. It also allowed me to come to know and become friends with many Turkana people, especially the members of four families, upon which much of my research is based. All have influenced the way that I think about human ecology and the way I have conducted my work. I hope that in some small way this book will make a contribution to the study of human-environmental relations, and a contribution to the literature on pastoral peoples.

It is impossible to thank all the people who contributed to the production of this book over the last twenty-three years. Hundreds, if not thousands, of people during this time have helped me refine my thoughts, debated issues, offered hospitality, taken care of me when I was sick, fixed my vehicle, and so on. I cannot list them all. Please forgive any omissions and realize that this is only a partial list of those who should be acknowledged.

Financial support for my research was secured from the National Science Foundation, the Norwegian Agency for International Development, the Social Science Research Council, and internal funds from the Binghamton University, the University of Georgia, and the University of Colorado. The Centre for African Area Studies at the University of Kyoto provided me with a summer fellowship which gave me the time and support to finalize the manuscript. I want to especially thank Dr. Masayoshi Shigeta for all his help in arranging this. Research permission was granted by the Office of the President of the Republic of Kenya. Without this support none of this research would have been possible.

Of course the people who really made this work possible were the Turkana people who let me into their lives and shared their joys and tragedies with me—made me feel welcome, and, at least to some extent, a part of their families. I cannot thank enough Angorot, Lorimet, Atot, Lopericho, and their families. It is also impossible to think of field research in Ngisonyoka without the help of Elliud Achwe and Lopeyon.

All members of the South Turkana Ecosystem Project contributed in some way to my research. Neville Dyson-Hudson helped conceptualize the project and included me in the early funding cycles. Both Neville and Rada Dyson-Hudson shared their wealth of knowledge of East African pastoralist people with me, and Rada and I shared research results and coauthored a book published by the Human Area Research Files in 1985. Mike Little provided intellectual and emotional support at times when I most needed it. Paul Leslie became a friend, colleague, and mentor in such diverse fields as human demography and behavioral ecology, as well as in motorcycle riding and repair. Layne Coppock, Kathy Galvin, Robin Reid, and Jan Wienpahl shared camp life, good conversation, and the trials and tribulations of conducting dissertation research with me. Jan Wienpahl shared the early days in South Turkana, as well as our breaks away from the field. The list of friends and colleagues who offered friendship and hospitality over the years is long, but Peggy Fry, Trevor Dixon, Peter Little, Dave Caddis, Diane Perlove, Randy Lintz, Jeanine Finnell, and the ever-changing complement of characters at Kuny Kastle stand out.

The late Jim Ellis was a friend and mentor in the field of arid land and range ecology. He influenced much of my thinking, and we continued to work together until his untimely death in 2002. Other ecologists, in particular Dave Swift and Mike Coughenour, shared time in the field and their understanding of ecosystem ecology and modeling.

Finally, I have to thank my family. My daughter, Kate, has had to accept that her father was often "in Africa." She has done this graciously and on two recent trips to Tanzania has become somewhat of a field-worker herself. Last, but certainly not least, I want to thank my wife, Judith. She is my emotional and intellectual partner. She has encouraged and shared aspects of my field research, and contributed in more ways than I can name in the production of this book.

PART 1

CHAPTER 1

Introduction

It seems to be taken for granted these days that a pastoral way of life is disappearing throughout the world. From the steppes of China to the savannas of Africa, the popular press portrays pastoral peoples as the last vestiges of a bygone age that will not survive the next one or two generations. Certainly pastoralists and pastoral livelihoods are changing; some people are becoming more sedentary, and livestock-based livelihoods are diversifying. But many of the stories that appear in the popular press are simplified and exaggerated, and not all pastoralists are settling down and diversifying their economies. The Turkana people of northwest Kenya are among the latter. They live in an ecological environment where the agricultural potential is extremely limited. They also live in a political environment that is very unstable where raiding and violence may erupt at any time. There are other people like the Turkana, in Uganda, Sudan, Ethiopia, and Somalia, to name a few places. How Turkana people cope with environmental and political instability articulates well with current discussions of climate change, arid lands as nonequilibrium-based ecosystems, and the proliferation of violence and war. They have much to teach us about coping with these stresses, about survival and resilience.

How pastoral peoples use the land and its natural resources, manage their livestock, and make decisions about where and when to move has been the subject of discussion for much of the twentieth century, and the debate continues today. The image of pastoral peoples, especially African pastoralists, as people engaged in a primitive and outdated livelihood wandering around in a bleak landscape and irrationally keeping large numbers of livestock primarily for prestige is long gone, although one can still see elements of this image in public pronouncements of some African politicians and embedded within development initiatives. Pastoral people are now viewed by many as keen decision makers, trying to cope with a difficult environment and a shrinking resource base. Among those who study pastoral peoples, mobility is seen as the major adaptive strategy that not only preserves the resource base, but allows for larger livestock holdings than would be possible were movements restricted or curtailed.

Just as our understanding of pastoral peoples has changed, so has our understanding of the relationship of people to the environments within which they live. Gone is the idea that environments determine how people will make a living or organize their social relationships. Most of those who study human-environment relationships accept the notion that all people live within ecosystems, but how ecosystems are structured and function is open to debate, as is the role of human populations within a given ecosystem.

Ecosystems and the "New Ecological Thinking"

The use of an ecosystem as a conceptual tool to help examine and explain human behavior, often often referred to as an "ecosystem approach," has helped advance our understanding of human-environmental relationships significantly (see chap. 2). However, it has also been subject to a number of criticisms, especially within the social sciences. One critique has focused on the nature of an ecosystem as a self-regulating entity where ecological relationships function to maintain equilibrium. Other critiques have targeted the lack of attention paid to economics, politics, and history as well as an underemphasis on the analysis of decision making at the individual and household level.

Indeed, what is now referred to as the "new ecological thinking" has become the study of "disturbance, disharmony, and chaos" (Worster 1990:3). According to geographer Karl Zimmerer the "new ecology" "accents disequilibria, instability, and even chaotic fluctuations in biophysical environments, both "natural" and "human impacted" (1994:108). Many natural and social scientists see the concepts of equilibrium and ecosystems as inexorably tied to one another. Worster, among others, sees the questioning of equilibrium-based models of the environment as a direct challenge to the ecosystem concept itself.

Some of these critiques may be muted by a new understanding of how ecosystems (at least some ecosystems) are structured and function, which began to emerge in the 1970s. The idea that highly variable ecosystems, such as arid-land ecosystems, were based on functional relationships that maintain equilibrium was questioned. The concept of persistent, but nonequilibrium-based ecosystems gained support over the last twenty-five years and has, to a large extent, replaced the equilibrium-based model for arid lands among natural scientists. The ecologists on the South Turkana Ecosystem Project (discussed later) share many of the views of those who emphasize instability and disequilibria, but have studied these forces within an ecosystem framework.

Regardless of whether one considers the new ecological thinking as a direct challenge to the ecosystem concept, it seems clear that few social scientists have embraced this idea and fewer yet have tried to operationalize the new thinking in the analysis of human-environmental relationships.

According to Zimmerer, "Our efforts in human geography to date have scarcely touched the profound re-interpretation of biophysical environments that has emerged through the perspective of the 'new ecology'" (1994:108). Writing more recently about the importance of this new thinking and its influence on social science, Scoones says:

> How have the social sciences attempted to articulate with ecological thinking over the past few decades? Too often, such social science analysis—whether in anthropology, sociology, geography, or economics—has remained attached to a static equilibrial view of ecology, despite the concerted challenges to such a view within ecology over many years. . . . The balance of nature has had a long shelf life in the social sciences, reinforced by functionalist models dependent on stable, equilibrial notions of social order. (1999:484)

A somewhat different perspective is developing within anthropology. Referred to by some as the "new ecological anthropology," the issues addressed "arise at the intersection of global, national, regional, and local systems, in a world characterized not only by clashing cultural models but also by failed states, regional wars, and increasing lawlessness" (Kottak 1999:26). This is in contrast to the "older ecologies," which were characterized by "the narrowness of their spatial and temporal horizons, their functionalist assumptions, and their apolitical character" (23).

Many aspects of the new thinking in understanding human-environmental relationships are relevant to the Turkana study considered in this book. A number of the key contributions to the development of arid lands as nonequilibrial ecosystems were made by ecologists working on the South Turkana Ecosystem Project, but they have yet to be utilized in our understanding of how individuals make decisions about resource use and mobility in their daily lives.

International, national, and regional events and policies also influence the daily lives of Turkana people and in the context considered here are reflected by increased levels of raiding and violence. Violence and raiding are not new to the Turkana people, or their neighbors, but the last two decades have seen some of the most devastating attacks on Turkana by neighboring groups in the last eighty years. Intertribal raiding is now a persistent threat, punctuated by short periods of extreme violence. Families can lose loved ones and their entire livestock holdings in a few hours or a few minutes. Understanding the reasons for this escalation of raiding is not easy or simple. It requires a political, economic, and historical perspective, in addition to an ecological one.

Both Scoones and Kottak emphasize the incorporation of scalar approaches, politics, and economics in the new ecological thinking. These are some of the same issues emphasized in much of the literature now referred to as political ecology, which has emerged as an important paradigm in some of the social sciences, especially geography. Although I am not framing this study within a political ecology context I do hope to demonstrate the importance of these issues in the account presented here. To a large extent I agree with Vayda and Walters (1999) in their critique of political ecology in that many publications in this field underemphasize the ecology, or lose it altogether. What I am trying to do here is to include these perspectives without sacrificing the basic ecological underpinning of the analysis.

GOALS AND THEMES

This book has three goals. The first is to bring the ecosystem concept back into the analysis of human-environmental relationships and to address some of the major criticisms of the ecosystem approach. I intend to do this by utilizing the concept of arid-land ecosystems as persistent but nonequilibrial systems, by focusing my analysis on the lives of individuals, and by examining raiding from ecological, historical, and political perspectives. The second goal is to demonstrate that an ecosystem approach and "other ecologies" are not necessarily antithetical. Of particular importance here are evolutionary, Darwinian, or behavioral ecology, with its emphasis on individual decision making and its focus on self-interest and reproduction, and political ecology, which emphasizes both a scalar approach and the importance of politics and history. The third goal is to provide some ethnographic richness to a study of human-environmental interactions, especially as this relates to the South Turkana Ecosystem Project. I have often felt that publications that explored human-environmental relationships were too dry; the humanity of the people who were at the center of the study was missing. Too often people became data, their behavior reduced to data sets. This book is in many ways a reaction to this; I include data and analyze data sets, but I hope that the people who are central to the study come across as individuals. Further, I hope that the reader comes to know the Turkana people that are the subject of this book, at least in a small way.

The major theme that frames discussions throughout the book is that an understanding of Turkana land use, livestock management, mobility, and decision making lies at the confluence of ecology, history, and politics. An examination of how the ecosystem is structured and functions, and how the human population articulates with the ecosystem, is necessary but not sufficient to understand the complexity of human-environmental relationships.

TURKANA AND THE SOUTH TURKANA ECOSYSTEM PROJECT

The Turkana are a pastoral people most of whom live in the arid and semiarid regions of the Turkana District in northwestern Kenya (map 1.1). They have subsisted throughout the twentieth century by raising five species of livestock: camels, cattle, sheep, goats, and donkeys. In

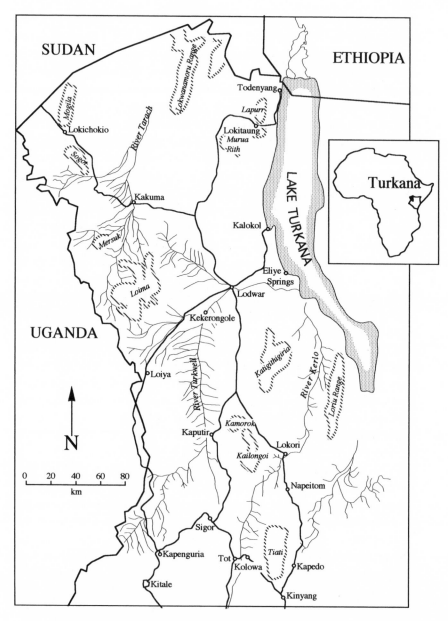

Map 1.1.   Turkana District. (Data from Soper 1985.)

some areas, their livestock-based economy is supplemented by the cultivation of sorghum and maize, and along the shores of Lake Turkana, some destitute Turkana pastoralists have taken up fishing. According to the latest census figures, their population currently exceeds 350,000, but this figure should be viewed as only a rough approximation.

For over twenty years I have been conducting research and writing about the Turkana people. Most of this work has been part of the South Turkana Ecosystem Project (STEP). This multidisciplinary project is one of the most detailed studies of a pastoral population ever conducted; it is also one of the most thorough examinations of a human group conducted within an ecosystems framework. The majority of the long-term multidisciplinary funding came from the National Science Foundation (NSF) and the National Institute of Mental Health (NIMH). These funds were supplemented by grants to individuals from a number of other agencies and organizations; especially important were the Wenner Gren Foundation, the National Geographic Society, and the Norwegian Agency for International Development (NORAD).

From 1980 to 1995 over 100 students, scientists, technicians, and other field-workers participated in data collection. The STEP literature includes fourteen dissertations, over two hundred articles and book chapters, and two books. The first book to bring together much of the anthropological literature was published in 1999 and synthesizes the major findings from the cultural and human biological components of the project (Little and Leslie 1999). Despite this record of publication, relatively little has been written about the lives of the individual Turkana people among whom this research has been conducted.[1]

Planning for the South Turkana Ecosystem Project began in the mid-1970s, and field studies were initiated in early 1980. Neville Dyson-Hudson, who had spent years working with East African pastoral peoples, primarily the Karimojong in Uganda, conceptualized the project. Although anthropologists had conducted many studies of pastoral peoples in the 1950s and 1960s that in many ways revolutionized our way of thinking about these peoples, Dyson-Hudson felt that knowledge concerning the environments within which pastoral peoples lived, especially how these environments were managed and exploited, was seriously deficient. He believed that it was critical to understand these arid ecosystems thoroughly and to study the live-stock population as well as the human population.

In the mid-1970s Dyson-Hudson initiated discussions with ecologists at the Natural Resource Ecology Laboratory (NREL) at Colorado State University to explore the possibility of launching a multidisciplinary project designed to examine the ecological and human ecological aspects of pastoral subsistence. Initially, two bodies of work influenced project thinking. The first was the cultural ecology approach in anthropology based on the works of Julian Steward and Daryl Forde. The other was based on the multidisciplinary research on human populations, which was given much credibility and impetus by the Human Adaptability and Ecosystem Analysis sections of the International Biological Program (IBP) during the 1960s (Little et al. 1990).

Ecologists from the NREL were approached because they were the principal scientists involved in the grasslands component of the IBP and were interested in rangeland studies internationally, as well as within the United States. The objectives of the overall project were to investigate (1) the effects of human resource extraction on the ecosystem, (2) patterns of human exploitation, herd management, and environmental effects on the social system, and the effects of each on the health and adaptability of the people (Little et al. 1990). The project brought together ecologists, social anthropologists, biomedical scientists, and human biologists from both Kenya and the United States.

It was decided that research would take place among the Ngisonyoka, a section of the Turkana ethnic group. The Turkana were chosen because they inhabit some of the most arid lands occupied by pastoralists in East Africa, and thus human-environmental relationships might possibly be more marked than among pastoral people living in less severe climates. The southern part of Turkana District had also been the research site of an expedition conducted by the Royal Geographic Society, and there were historical environmental and anthropological data to draw upon. Dyson-Hudson participated in the Royal Geographic Society's expedition and thus had contacts among the pastoralists living there.

The basic research was organized around four broad themes that Little and Leslie (1999) articulated in the form of questions.

Sociocultural Anthropology. How do the Turkana extract resources from, survive in, and adapt to an arid and stressful environment by nomadic pastoral subsistence?

Ecology. How do these extractive techniques, patterns of resource
   exploitation, and livestock modify the dry savanna ecosystem?
Human Population Biology. What effects do the environment and
   sociocultural practices have on health and adaptability of the people?
Demography. How do the social conditions, ecology, and human biology of the Turkana contribute to the maintenance and persistence
   of the human and livestock populations? (1999:15)

Summary of Major Findings and a Guide to the Literature

Ecology

Early ecological research was devoted to the development of models
of arid and semiarid ecosystems, an understanding of livestock feeding ecology, and the analysis of energy flow. Jim Ellis was responsible
for the overall project design of the ecological research. With the rare
ability to envision the big picture, he remained the intellectual leader
of the ecosystem studies throughout the entire project. The modeling
effort begun by David Swift and built upon by Michael Coughenour
eventually led to the construction of the Savanna model (Coughenour
1991, 1992), one of the most complex ecosystem models ever developed and now used around the world in the management and conservation of natural resources.

Layne Coppock's work utilized methods more common in wildlife
ecology than for domesticated livestock and broke new ground in
understanding how the five different species of livestock herded by
the Turkana utilized the vegetal resources available to them (1985;
also see the appendix for a list of numerous articles by this author).
The examination of energy flows incorporated Coppock's findings as
well as Galvin's work on human nutrition and McCabe's research on
human and livestock mobility. This resulted in an article published in
Science (Coughenour et al. 1985) that traced the transformation of
energy from the vegetation, through the livestock, and to humans.
The analysis of ecosystem dynamics also concluded that Turkana pastoralism did not degrade the environment, contrary to the commonly
held belief that nomadic pastoralism was an inherently destructive
land-use practice (see Lamprey 1983). Further ecological studies by

Mugambi (1989) on dwarf shrub ecology, Patton (1992) on landscape patterning, and Reid (1992) on the influence of pastoral management strategies on the growth of trees added to our understanding of the structure and dynamics of the Turkana ecosystem.

In 1988 Ellis and Swift published what became a seminal article in the understanding of the ecology and development alternatives for arid lands. In "Stability of African pastoral ecosystems: Alternate paradigms and implications for development" (1988), the two senior STEP ecologists laid out the argument for conceptualizing arid lands as persistent but nonequilibrium ecosystems. This was expanded in a book chapter by Ellis, Coughenour, and Swift (1993). I discuss this issue at length later in the book and so will not expand on this concept here.

Sociocultural Anthropology

The early cultural anthropology studies built upon Dyson-Hudson's notion that an examination of livestock characteristics and livestock management was critical to understanding major features of Turkana social organization. In addition, the roles of small stock in pastoral livelihoods and of women in the management of resources were emerging as important themes in the larger literature. Thus McCabe focused his initial research on livestock management and mobility (1985; also see the appendix for a list of numerous articles by this author), while Wienpahl (1984a) concentrated on the contribution that goats and sheep made to overall subsistence and the management and dynamics of the flocks. As will be seen later, it was impossible to ignore the impact of raiding in the management of livestock, and this is reflected in both McCabe's and Wienpahl's dissertations, as well as the book by Dyson-Hudson and McCabe (1985) published by the Human Relations Area Files. Later researchers again took up the theme of raiding and violence based on their experiences in the 1990s (Gray et al. 2003).

The extent to which social networks formed a social safety net was recognized as critically important to the long-term viability of families in a highly variable environment, and Johnson (1990) concentrated on an examination of social networks in his dissertation research. Both Neville and Rada Dyson-Hudson conducted research throughout the South Turkana Ecosystem Project, taking a broad perspective on how pastoralists adapt to arid environments (N. Dyson-Hudson 1985, 1991; R. Dyson-Hudson 1980, 1983a, 1983b, 1989). As the project progressed it became clear that many Turkana pastoralists failed and dropped out of the pastoral sector, often leaving the district entirely. Rada Dyson-Hudson's more recent work examining demographic data revealed that 49 percent of ethnic Turkana were living outside of Turkana District, most engaged in subsistence activities other than pastoralism (R. Dyson-Hudson and Meekers 1999).

The cultural anthropological studies resulting from the South Turkana Ecosystem Project provide one of the most detailed pictures of life among a pastoral people and the most comprehensive examinations of the interrelationship among environmental variables and aspects of human behavior ever conducted among a pastoral people. Turkana social organization was found to be flexible, and individual responses to environmental stress highly variable among individuals. Families and herds were extremely mobile, but raiding and the threat of violence exerted a strong influence on decision making and the exploitation of natural resources (this volume).

Human Biology

From the very outset of the South Turkana Ecosystem Project it was understood that the health and adaptability of individual Turkana people would serve as a "measure of the society's ability to support itself " (Little and Leslie 1999:19). Six categories of variables were identified that could measure the health and adaptability of Turkana individuals: diet and nutrition; infant, child, and adolescent growth; adult size and body composition; activity and physical fitness; reproduction; and disease (21).

Diet and nutrition were studied initially by Galvin (Galvin 1985, 1988; Galvin and Little 1999) and later by Gray (1994b, 1998). Infant and child growth was the subject of Gray's dissertation (1992) and resulted in numerous subsequent articles (see appendix). Michael Little led the study of human biology and examined many aspects of growth and development among all age and sex categories of individuals studied (Little and Leslie 1999; also see the appendix for a list of numerous articles by Little). Pike's research on women's health and pregnancy outcomes articulates well with both Gray's and Little's work (Pike 1995). Body composition and fitness levels were examined in the early stages of the project by Galvin (1985) and Little (Little, Galvin, and Mugambi 1983; Little, Galvin, and Leslie 1988), and later by both Little and Gray (Little and Gray 1990; Little, Gray, and Leslie 1993). Curran's dissertation research on work capacity examines the ability of individuals to perform certain physical activities given the constraints of low caloric intake, low body fat, and high temperatures (Curran-Everett 1990; Curran and Galvin 1999). Disease affects individuals of all age and sex categories and impacts people's ability to perform certain activities, reproduce, and survive. Disease is physical, but it is also culturally constructed. Shell-Duncan examined morbidity among Turkana children (Shell-Duncan 1994, 1995), while Shelley focused on how Turkana conceptualized disease and their indigenous system of health care (1985).

Overall, the human biology studies revealed that the nomadic Turkana were quite healthy, despite an extremely low level of caloric intake. Low body fat leaves little reserves, and respiratory diseases, measles, and malaria can have devastating effects especially on children. Individuals grow more slowly than their Western counterparts and do not achieve the height or weight norms set out by the World Health Organization; however, they continue to grow into their young adult years.

The ability to adapt to the stresses of an arid, hot, and variable environment required both biological and behavioral responses. Behavioral responses could be seen at two levels: on a daily basis, such as structuring day-to-day activities to minimize the stresses associated with heat and dehydration (Curran-Everett and Galvin 1999), and at the cultural level, whereby resources are shared and activities allocated according to age and sex, within, as well as outside of, the homestead.

Demography and Reproduction

By the second year of fieldwork, it became clear that an examination of human population dynamics was necessary to understand the relationship between the environment and people. The ability of the human population to properly manage livestock depends on a large enough labor pool, which in turn depends on the size and composition of the family, as well as rates of fertility and mortality. At the time that the South Turkana Ecosystem Project began, there were very few studies of pastoral peoples, and this alone would have justified engaging in this component of the overall research (Little and Leslie 1999:18). Initial research (Brainard 1981, 1982) suggested that the nomadic Turkana had higher fertility rates than those who had settled, and this was contrary to what was known about other pastoral populations, further emphasizing the need for demographic research. Anthropological demographers Paul Leslie and Peggy Fry joined the South Turkana Ecosystem Project in 1981, and Leslie directed the demographic studies throughout the project.

Leslie felt that a biobehavioral approach to the demographic work was necessary, and the demographic studies would "serve as a bridge between the research in sociocultural anthropology and in human population biology" (Little and Leslie 1999:19). The initial focus of the demographic research was to define fertility rates. Later research addressed the seasonal fluctuations in fertility and the complex relationships among "breast feeding, subsistence patterns, birth spacing, infant morbidity and mortality, and maternal health" (19). Studies of pregnancy loss, female reproductive function, and later, male reproductive function were undertaken in the final stages of the project.

Benjamin Campbell and Kenneth Campbell (no relation) joined Leslie in the later demographic research (Leslie and B. Campbell 1995; Leslie, B. Campbell, and Little 1993; Leslie, K. Campbell, and Little 1993). The demographic analysis was closely articulated with the research conducted by Rada Dyson-Hudson, Sandra Gray, and Ivy Pike (mentioned earlier). Toward the end of the project Michael DeLuca examined pregnancy loss among settled Turkana in the town of Morulem.

The demographic research revealed that nomadic Turkana women have an average of 6.6 live births and that their fertility is "high, but not remarkably so" (Leslie et al. 1999:277). The seasonal distribution of births was significant, and in fact is one of the most seasonally skewed patterns ever recorded for a human population (Leslie and Fry 1989; Leslie et al. 1999). The comparison of settled and nomadic women showed that the fertility of the nomadic women was indeed higher than the settled women, and that settled women have a higher rate of intrauterine mortality (Leslie et al. 1999:277–78).

The South Turkana Ecosystem Project remains the most comprehensive multidisciplinary project that has involved humans within an ecosystem framework. It is impossible to do justice to the integrated studies or to the importance of individual research projects in this summary. What I am writing here is intended to be a guide for those readers wishing to delve deeper into that project. Little and Leslie's edited book expands on many of the issues summarized here and is a significant resource, especially for those interested in the sociocultural, human biological, and demographic components of the South Turkana Ecosystem Project. I have included an appendix of publications that have emerged from the Turkana project at the end of this volume, as I have only mentioned a few individual articles and book chapters here to illustrate and support the points mentioned previously.

Fieldwork

My own research among the Turkana people began in 1980, and I last visited the area in 1996. During this sixteen-year period, I spent forty-eight months living in southern Turkana, examining issues related to herd management, decision making, patterns of human and livestock mobility, response to and impact of drought, and the influence of development projects on indigenous systems of coping with environmental stress. Although the South Turkana Ecosystem Project had as a unifying theoretical perspective the understanding of the dynamics of an arid land ecosystem and the role that humans play in that ecosystem, I have focused much of my research on how four individual families have coped with an extremely variable and harsh environment.

In February 1980, I made my first visit to Turkana District accompanied by Michael Lokoruka, one of only two Turkana individuals who had graduated from a university and who had worked with David and Judith MacDougall on a series of Turkana films. Also with us was Joseph Lomuria, at the time a university student and son of a local politician from the town of Lokori. In May, Jan Wienpahl, a student of Robert Netting, came to Kenya to work with me on the Turkana project. By then initial contacts with government representatives had been made and a vehicle purchased. In June, Jan and I drove to Lokori, where we were to begin our research, and made preparations for a three-week walking safari to meet Turkana pastoralists. We were told that it would be difficult to travel with a vehicle in the bush and that it was preferable to walk and use donkeys for pack animals.

We needed donkeys, someone who knew how to take care of them, and someone to act as a guide and interpreter. We found that guide and interpreter in Elliud Achwee Lowoto, a young Turkana man who had worked in the mission dispensary. Elliud was the son of a local pastoralist and one of the few children from a pastoral family who had attended school. Because he had attended primary school he spoke some English. We also hired two other men, Julius and Lopeyon, who would help with camp and take care of the donkeys. Lopeyon had been an itinerant trader and had traveled all over southern Turkana exchanging tobacco and other goods for livestock. Julius was half Samburu and half Turkana and said that he knew both peoples equally well.

As the sun was setting on the evening of July 21, after a very hot and searing day, Jan and I set out, accompanied by Elliud, Julius, Lopeyon, and five donkeys purchased in Lokori. We wanted to meet with a man called Lopeii, who had worked with Neville Dyson-Hudson previously, but other than this we had little control over where we were going from one day to the next. In fact, during the entire journey, we never knew where we were until the outline of the shops in Lokori appeared on the horizon on our return three weeks later.

Jan and I had decided to concentrate our research on a few families and try to get to know them well. The advantage of this was tighter control of data collection, but the disadvantage was working with a small sample, which limited the representativeness of our conclusions. We did meet Lopeii but found him to be aggressive and greedy. He was also wealthy and expected us to do far more for him than we thought we would be able to do. Fortunately we also met many other Turkana families and eventually settled on four herd owners and their families with whom we thought we could work and who seemed to be willing to put up with us. These herd owners were Angorot, Lopericho, Lorimet, and Atot. Each family differed in size and livestock wealth, although there were no very wealthy or extremely poor people in our initial study. These four families became the focus of our research, and their land-use practices and livestock eventually became integral components of the ecological research conducted by the STEP ecologists.

Although I interacted with all family members, I concentrated on trying to understand the process by which the herd owner made decisions concerning movement and the management of livestock. Thus, in this book the emphasis is on the land-use practices of the four herd owners that I met in early 1980. My understanding of Turkana life was gained primarily from these men.

During the following decade, I made nine separate trips to Turkana District for a total of forty-six months spent in the field. I returned again in 1996 for two months to catch up on events that had transpired since my last visit in 1990. Each trip had a different research focus, but I was always able to visit with members of the four families and tried to incorporate them into the research whenever possible.

STRUCTURE OF THE BOOK

The book is divided into four parts. Part 1 reviews the relevant literature and summarizes what is known about the ecology of Turkana District and the Ngisonyoka section. Chapter 2 discusses the ecological and African pastoralist literature that partially frames the case study. As one of my goals in writing this book is to examine the extent to which pastoralists conform to the new ecological thinking of arid lands as nonequilibrium ecosystems, I have included a rather detailed history of the role of ecosystem analysis in anthropology and the emergence of the paradigm of nonequilibrial ecosystems. I also discuss other ecologies (behavioral ecology and political ecology) and the way individuals have been incorporated into or left out of research in ecological anthropology. In chapter 3, a general overview of the environment of Turkana District is presented, as well as a brief summary of the history and social organization of the Turkana people. Chapter 4 concentrates on the ecology of the Ngisonyoka section of the Turkana and describes how the Ngisonyoka conceptualize their landscape. Chapter 5 looks at the larger literature on tribal warfare and raiding to help frame the case study and the current issues being debated. I follow this with a perspective on the literature on raiding and violence among the pastoral peoples in East Africa, ending with a discussion on how this has specifically impacted the Turkana people.

The case study in part 2 includes a description of the four families of Angorot, Atot, Lorimet, and Lopericho and their herds. Chapter 6 presents a detailed account of these four families whom I came to know intimately over time. In chapter 7 I present a somewhat detailed narrative of the movements and decisions that were related to mobility and herd management for each of the families. I have included material about the environmental and political context for each season and try to illustrate how this influenced the decisions that were made by the herd owners.

In part 3, analyses of individual-level movements, decision making, and the impact of these decisions on both the family and the herds are presented. Chapter 8 focuses on the analysis of mobility to identify patterning where it exists and explain variability where it does not. In chapter 9, I examine how the growth of herds and that of the family are interrelated, and how the two are intertwined in the decision-making process.

Part 4 shifts the analysis from the level of the individual to that of the group, and concludes the study with a discussion of major findings and their relevance to theory and policy. In chapter 10, I examine the movements and herd dynamics of four sections of the Turkana that were the subject of a yearlong study on the impact of drought, conducted by myself and members of the STEP team. I do this so that the detailed data on the Ngisonyoka can be contextualized within a larger framework. Here I am looking for patterns that may be obscured in the analysis of individuals. I am also looking for regularities that are found across sections. In chapter 11, I bring the discussion back to issues and questions raised in the preceding chapters and also provide an update on more recent events that have impacted the Turkana people. I occasionally insert vignettes that help the reader understand events that stood out for me as particularly important to the research or my recording of it.

CHAPTER 2

Review of the Literature and Theoretical Framework

This chapter presents an overview of the ecological and African pastoral literature that partially frames the case study and theoretical arguments considered later in the book. Almost all of my research among the Turkana people was conducted as part of the South Turkana Ecosystem Project, therefore a review of the literature relating to ecosystems and anthropology is a logical place to start. Some of the most important contributions to the "new ecological thinking" (Scoones 1999) that view arid and semiarid lands as persistent but nonequilibrium systems were conceptualized and developed by the ecologists on the STEP. Because one of my objectives in writing this book was to explore the extent to which the actual behavior of the individual Turkana people with whom we worked corresponded to that predicted by this new understanding of ecosystem dynamics, I have tried to follow the development of this theoretical approach from its beginnings to the issues being debated today.

The book, however, is not a defense of ecosystem studies, and as will be evident later, many of the decisions relating to natural resource use and much of the observed behavior are better understood as responses to political events and uncertainties than to the ecology. This fits more closely with the approach championed by political ecologists than rangeland ecologists. Decisions about livestock management and the transfer of livestock among families are impossible to understand without considering the relationships between herd growth and family formation and development. Because much of the important literature here has been published in the field of evolutionary or behavioral ecology, I have provided the reader with a brief overview of these other ecologies.

Finally, the book is about pastoralism, specifically about African pastoralism. I conclude this chapter with an overview of pastoral land use and discuss how an ecological explanation for mobility, livestock management, and decision making has been both utilized and critiqued. As mentioned in the introduction, what is presented here is only a partial review of the literature that frames the case study. The literature on tribal warfare and raiding and on warfare among East African pastoral peoples is reviewed in chapter 5.

ECOSYSTEMS AND ANTHROPOLOGY

The concept of the ecosystem was introduced by Sir Arthur Tansley in 1935 to the ecological community. Almost thirty years later the concept was introduced into the anthropological literature with the publication of *Agricultural Involution* by Clifford Geertz (1963). Although the examination of human-environmental relationships had been a subject of serious study in anthropology for much of the discipline's history, the adoption of concepts such as ecosystem as an analytic unit marked a significant shift in how ecological anthropologists perceived and studied human populations and their environment.

Tansley first presented the ecosystem concept in a discussion concerning the notion of the biological community and the idea that communities would change through time until a "climax" community was reached (Golly 1984). Tansley felt that the biological organisms and the physical environment in which they lived were far more interrelated than was assumed by the studies of communities common at that time. The focus of these community studies was on how the environment affected vegetation, organisms, or populations, usually within the context of successional change through time. Much of the understanding of community ecology and successional change was based on the work of F. E. Clements (1916).

Central to Tansley's notion of the ecosystem was the idea that this web of interrelationships produced a state of equilibrium, which fit well with naturalists' ideas concerning the balance of nature (McIntosh 1985; Moran 1990). The ecosystem concept gained gradual acceptance throughout the 1940s and 1950s, and the publication of Odum's *Fundamentals of Ecology* marked its place in the mainstream of ecological thinking. During the 1960s ecological research became more systems oriented with "ecoenergetics the core of ecosystem analysis" (Odum, quoted in McIntosh). This "new ecology," with an emphasis on energy flow, equilibrium, and self-regulation, is the paradigm that influenced the anthropologists who first began to use the ecosystem concept in their own work.

In early ecosystems studies conducted by ecologists, the role of humans was usually relegated to a form of disturbance or an external factor in the natural order. Anthropologists sought to place humans at the center of ecosystems studies. Although Geertz is credited with the first use of the ecosystem concept in anthropology, the publication of Rappaport's *Pigs for the Ancestors* (1968) demonstrated the potential explanatory power of ecosystem studies in the analysis of human behavior. The use of the ecosystem as an analytic unit was a significant departure from the way that human-environmental relations were studied in the 1950s and 1960s.

At the time, Julian Steward's (1955) cultural ecology was viewed as the dominant paradigm in understanding human-environmental relationships. Steward was not concerned with the environment per se, but with critical food resources and the manner in which they were exploited. He argued that the natural resources directly connected to the subsistence system constituted the "effective environment," and the human behaviors that were involved in the use of the these resources constituted the cultural "core." The manner in which resources were exploited would have a strong influence upon core forms of social organization, and these social forms would have an influence upon many other aspects of people's lives and social organization.

Steward viewed humans as unique, and he made no attempt to connect his research with that being conducted by ecologists or biologists. His influence has been profound in anthropology, and his ideas concerning the importance of understanding the impact of subsistence strategies to social organization can be found among many anthropologists working today, including myself. Despite the theoretical advances made by Steward, his "cultural ecology" was criticized for ignoring important environmental variables (such as disease and population pressure) and for being subjective in identifying aspects of the "effective environment" and the "culture core" (Netting 1968; Ellen 1982).

The difference between Steward's approach and the new ecological anthropologists' approach was emphasized in an article published by Vayda and Rappaport (1968), in which they argued that the focus on humans as unique was inappropriate and that human-environmental relations could only be really understood by viewing humans as part of larger ecological systems. They called for a unified field of ecology including humans and advocated that anthropologists begin to use the terminology and analytic units used by ecologists: individuals, populations, and ecosystems.

Following the lead of Vayda and Rappaport, a cadre of committed ecological anthropologists set out to adopt, to the extent that they could, the ecosystem as an analytical unit, with humans being one of many species involved in a system of self-regulation. The call for research on the role of ecoenergetics made by Odum for ecologists was also adopted by these ecological anthropologists

During the early 1970s a number of studies were conducted focusing on the flow of energy through the human population, prompting some critics to label this type of research a "caloric obsession" (for examples of this research see Kemp 1971; Rappaport 1971; Thomas 1976). In addition to energy flow studies, a number of ecologically oriented anthropologists set out to explain what they perceived as ecological puzzles found among some peoples throughout the world. These ecological puzzles were often religious beliefs or rituals that prompted people to act in a manner that seemed contrary to what one would expect within an ecological, self-regulating ecosystem.

By the mid-1970s, however, this type of ecological anthropology was criticized as being on the one hand reductionistic and on the other merely a new form of functionalism—and a naive form at that (see Friedman 1974; Ellen 1982). The criticism of the "new ecological functionalism" and the perception of ecosystems as closed, self-regulating systems was so strong within the anthropological community that by the 1980s ecosystem studies in cultural anthropology became increasingly rare.

Ecosystem analysis continued to develop within ecology. Ecosystem studies as conducted by cultural anthropologists were of a very different nature from those conducted by ecologists. Ecosystem studies in ecological research were conducted by well-funded multidisciplinary teams of ecologists and biologists. Ecosystem studies in cultural anthropology usually involved a single social scientist trying to fit his or her data into an ecological framework using whatever published material was available. The one exception to this was the incorporation of human biologists and physical anthropologists into the multidisciplinary ecosystem studies conducted within the International Biological Program (IBP). Michael Little, who participated in the IBP studies, has pointed out that differences in conceptual frameworks and terminology hampered the interchange of ideas among the social scientists and the ecologists. In addition, problems concerning how to define boundaries when humans are incorporated into ecosystem analysis and the extreme complexity involved severely limited progress in bridging the gap between anthropology and ecology within an ecosystems framework (Little et al. 1990).

In 1982 Emilio Moran organized a session at the American Association for the Advancement of Science Meetings to consider the contributions made by, and the limitations of, the ecosystem concept to anthropology. While most of the contributors to this session felt that the ecosystem concept was a useful heuristic device that encouraged quantification and systemic holistic thinking, they also felt that there were some serious limitations to the application of the ecosystem concept to the study of human behavior. In the introductory chapter to a volume published in 1984, based on the results of the 1982 AAAS session, Moran summarized what the participants saw as the major contributions and limitations of the ecosystem concept to anthropology.

In terms of the contribution Moran stated:

Just as the ecosystem concept helped biology broaden its interests to include the neglected physical environmental factors, so it affected anthropology. The ecosystem approach provided greater context and holism to the study of human society by its emphasis on the biological basis of productivity and served as a needed complement to the cultural ecology approach. By stressing the complex links of mutual causality, the ecosystem concept contributed to the demise of environmental and cultural deterministic approaches in anthropology. (1984:13)

The limitations were summarized as follows: (1) a tendency to reify the ecosystem and give it the properties of a biological organism; (2) an overemphasis on predetermined measures of adaptation such as energetic efficiency; (3) a tendency for models to ignore time and structural change, thereby overemphasizing stability in ecosystems; (4) lack of criteria for boundary definition; (5) level shifting between field study and analysis.[1] In a substantial revision and updating of the book published in 1990, Moran lists the same limitations with one addition: the tendency to neglect the role of individuals.

ECOSYSTEMS AS NONEQUILIBRIUM SYSTEMS

During the late 1970s some ecologists began to question the basic assumptions upon which much of the anthropological use of the ecosystem concept was structured, that of equilibrium. Holling (1973), for example, felt that natural systems are continually in a "transient state" and argued that emphasis should be placed on the conditions that lead to their persistence. The ability of a ecological system to withstand trauma played a central role in Holling's conceptualization. Among ecological anthropologists, Andrew Vayda and Bonnie McCay were the first to adopt Holling's emphasis on stability, persistence, and resilience and also provided an early critique of equilibrial ecological systems (1975). Vayda and McCay felt that ecological anthropologists would be better served by studying how ecological and human communities responded to natural hazards and thus were persistent, rather than work with the paradigm of a equilibrium-based, self-regulating ecosystem.

According to Worster (1990), it was the 1973 publication of an article by William Drury and Ian Nesbit that provided one of the most important challenges to the ecosystem concept based on the notion of equilibrium. In this article Drury and Nesbit argued that forests in the northeastern United States did not follow a pattern of succession as advocated by Clements. They found no evidence of progressive development and none of the criteria that Odum had identified as characteristic of mature forests. The attack on the idea of succession was also an attack on the notion that mature ecosystems were based on equilibrium relationships. Worster views this as the most important early challenge to the ecosystem concept itself.

An additional challenge to an equilibrium-based understanding of ecosystems came from researchers working in arid and semiarid lands. In 1973, Noy-Mier proposed that dry ecosystems are controlled more by climate than by biotic interactions. However, the article upon which many of the nonequilibrial arguments have been built was published by John Wiens in 1977. Wiens's research, focused on avian communities, found that predictions concerning community structure based on equilibrium theory began to break down in arid and semiarid environments. In future publications he went on to propose that in arid environments interactions among species are "decoupled" and that "species should respond to environmental variations largely independently of one another" (1984:451). He also felt that population dynamics in arid environments were strongly influenced by "abiotic agents" and not by species density or limitations of resources.

Wiens proposed that natural communities exist along a continuum from equilibrium states to nonequilibrium states; equilibrium communities would be expected to develop in areas of low to moderate environmental variation, whereas environments that are characterized by harsh and unpredictable climatic events would develop nonequilibrium communities (Wiens 1977).

In 1988, Jim Ellis and David Swift, ecologists working on the South Turkana Ecosystem Project, published a summary of the results of their research to date and found that the structure and dynamics of the South Turkana ecosystem corresponded closely with the model of nonequilibrium systems proposed by Wiens (1988). They found that climate exerted a strong influence on the vegetation, with only weak linkages between the herbivore population and the plant community and condition. The major perturbation was drought, and they noted that rainfall had dropped 33 percent or more thirteen times in the previous fifty years. Livestock condition and production closely tracked seasonal plant production, and during droughts livestock losses of over 50 percent were common. They summarized their results as follows.

The results of our work in Ngisonyoka Turkana . . . reveal anything but an equilibrial ecosystem. Here in the arid northwest corner of Kenya, pastoralists are locked in a constant battle against the vagaries of nature and the depredations of neighboring tribesmen. . . . However, despite the dynamic nature of the ecosystem, there is little evidence of degradation or of imminent system failure. Instead, this ecosystem and its pastoral inhabitants are relatively stable in response to the major stresses on the system, e.g., frequent and severe droughts. In theoretical terms this is a non-equilibrial, but persistent ecosystem. (1988:453)

Ellis and Swift went on to identify characteristics of equilibrial and nonequilibrial ecosystems, much as Wiens had done for communities. Their characterizations are presented in table 2.1.

In 1993, Ellis and Swift added a modeling dimension to their analysis and reconfirmed their previous findings concerning the nonequilibrial nature of arid land ecosystems (Ellis, Coughenour, and Swift 1993). They also mention support for their ideas coming from research conducted in the arid rangelands of Australia (Caughley, Shepherd, and Short 1987). Caughley and his colleagues also found that the characteristics of arid ecosystems were best explained as nonequilibrial systems and proposed that these types of systems will be found in areas where the coefficient of variance for rainfall exceeds 30 percent.

This perception of arid and semiarid lands as persistent, but nonequilibrial ecosystems has gained increasing acceptance among rangeland ecologists since the publication of Ellis and Swift's 1988 article (see Behnke and Scoones 1992; Behnke, Scoones, and Kervin 1993; Scoones 1996). An understanding of ecosystems (at least some ecosystems) as nonequilibrium systems mutes many of the major criticisms voiced by ecologists and anthropologists, especially those concerning time and structural change and those concerning the impact of extreme environmental events.

Finally, the notion that equilibrium exists at all, in what Berkes, Colding, and Folke (2003) refer to as "coupled social-ecological systems," is currently being challenged. According to these authors, "It is evident that the dominant worldview in resource and environmental management of 'systems in equilibrium' is incompatible with observations of the complex dynamics of social and ecological systems" (2003, xi). Like the "new" thinking in ecological anthropology and ecology, theory put forward by researchers working within the framework of coupled social-ecological systems emphasizes uncertainty, unpredictability, scalar approaches, and the incorporation of politics, economics, and so on into the understanding of human/environmental relationships. However, stress is also placed on the concepts of nonlinearity, complexity, self organization, and the possibility that systems may organize around a number of equilibria or stable states (detailed discussion of these ideas can be found in Berkes, Colding, and Folke 2003 and Gunderson and Holling 2002).

TABLE 2.1.   Characteristics of Equilibrial and Nonequilibrial Grazing Systems

	Equilibrial Grazing Systems	Nonequilibrial Grazing Systems
Abiotic patterns	abiotic conditions relatively constant	stochastic/variable conditions
	plant growing conditions relatively invariant	variable plant growing conditions
Plant-herbivore interactions	tight coupling of interactions	weak coupling of interactions
	feedback control	abiotic control
	herbivore control of plant biomass	plant biomass abiotically controlled
Population patterns	density dependence	density independence
	populations track carrying capacity	carrying capacity too dynamic for close population tracking
	limit cycles	abiotically driven cycles
Community/ ecosystem characteristics	competitive structuring of communities	competition not expressed
		limited spatial extent spatially extensive
	self-controlled systems	externalities critical to systems dynamics

Source: Data from Ellis and Swift 1988.

These ideas have had a major impact on the way some researchers and development planners view the future of arid rangelands, in particular the arid and semiarid rangelands and savannas of Africa. Those who view arid rangelands as persistent but nonequilibrial ecosystems tend to support what is referred to as the alternative view, as opposed to the mainstream view (see conclusion).

OTHER ECOLOGIES AND THE ROLE OF INDIVIDUALS

The role of individuals in ecosystem-level analysis within the social sciences has been all but ignored (as noted earlier). The emphasis on group-level dynamics and system function has marginalized both the role of individuals as analytic units and the impact of system-level processes on the behavior of individuals. Commenting on the literature before the mid-1980s, Rappaport has said:

> Earlier analysis in ecological anthropology . . . did not pay sufficient attention to the purposes motivating individual actions, actions which when aggregated constitute group events; nor did they pay sufficient attention to the behavioral variations among individuals, to differences in the understandings of the world entertained by individuals, to individuals as adaptive units, or to the conflicts between individual actors or between individuals and the groups to which they belonged. (1990:62)

Behavioral Ecology

The one area within ecological anthropology in which the analysis of individual-level behavior plays a critical role is that referred to as evolutionary ecology, behavioral ecology, or Darwinian ecological anthropology (for examples, see Cashdan 1990; Smith and Winterhalder 1992; Borgerhoff-Mulder and Sellen 1994; for a critique, see Vayda 1995). Behavioral ecology within anthropology began to coalesce as a field about twenty years ago and now has a number of strong adherents as well as critics. The unifying theme within human behavioral ecology is that human behavior is best understood within an evolutionary framework, and because adaptation is seen as occurring at the level of the individual rather than the group, the basic analytic unit should be the individual. Aggregate behavior of individuals can then be used to understand behavior at the group or population level.

Anthropologists working within the human behavioral ecology framework usually combine "methodological individualism" with rational choice theory to produce models of behaviors that emphasize the optimization or maximization of some "currency," most commonly reproductive success (but food intake or energy capture are often used as proxy currencies). Smith and Winterhalder describe the role of methodological individualism as follows.

> Methodological individualism (MI) holds that the properties of groups (social institutions, populations, societies, economies, etc.) are the result of the actions of its individual members. . . . The primary role of MI is to provide "microfoundations" or "actor based accounts" for social phenomena by analyzing the extent to which they are the aggregate outcomes of individual beliefs, preferences, and actions. (1992:39)

Much of this body of work examines decision making among hunting and foraging peoples, but some anthropologists have also applied this perspective to the analysis of decision making among pastoral peoples. Probably the best-known studies are those that use behavioral ecology to examine the relationship of wealth and reproductive success in pastoral societies (Irons 1979; Borgerhoff-Mulder 1988, 1992, 1996; Cronk 1989). Others use this paradigm to examine the rationale for pastoralists manipulating the species structure and reproductive pattern of pastoral herds (Mace 1990; Mace and Houston 1989). To my knowledge only one attempt has been made to apply a behavioral ecology perspective to livestock movements, and that has involved herd management strategies of sedentary livestock keepers (Deboer and Prins 1989).

One of the most widely accepted critiques of this theoretical perspective comes from those who view the lack of attention paid to culture to be as egregious as ignoring the biological (Rappaport 1990; Vayda 1995). Some human evolutionary ecologists are trying to grapple with the inclusion of culture in their analysis as seen in the work of Boyd and Richerson (1992) and Cronk (1995).

Although these perspectives have taken very different approaches to the study (or lack of it) of individuals, some anthropologists (including myself) do not feel that the study of individual behavior within a larger systems framework is impossible. Andrew Vayda with his method of "progressive contextualization" has been advocating this for years. Progressive contextualization involves "focusing on significant human activities or people-environment interactions and then explaining within progressively wider or denser contexts" (Vayda 1983:265). Although Vayda has stressed the importance of using both individuals and groups in ecological analysis, he has also criticized the use of the ecosystem (Vayda and McCay 1975) and the use of human behavioral ecology, which he refers to as Darwinian ecological anthropology (Vayda 1995).

Many social scientists that find the ecosystem approach useful also recognize the importance of individual-level analysis. Butzer, for example, feels that the analysis of individual decision making and aggregate human behavior can and should be directly incorporated into ecosystem analysis (1990). Moran calls for future research in ecological anthropology to integrate "multiple models, focused at different levels of detail" (1990).[2] Rappaport has stated that "the relationship of actions formulated in terms of meaning to the systems constituted by natural law within which they occur is the *essential problematic of ecological anthropology* (1990:67, emphasis in original). He goes on to say that "the banishment of any one of these levels—the individual, the social, the ecological—from our analyses is a very serious mistake, conferring upon us no benefits and costing us dearly" (68). In many respects the theoretical focus of this book can be seen as a response to these calls for multiple models and the integration of individual- and systems-level analysis.

Political Ecology

The final ecology to consider is political. Political ecology grew out of a response to analyses of human-environmental research that was considered to be too reductionistic and too synchronic. Although the term had been used in the social sciences for some time (it was first introduced into the literature by Eric Wolf in 1972), Blaikie and Brookfield are usually credited with elevating "political ecology" into an emerging paradigm for human-environmental research. Their concern was to combine ecological analysis with that of political economy with a focus on the use and management of natural resources at the local level. Political ecology incorporates historical, political, and economic analyses within a scalar framework. How resources are used and the institutions that govern that use are examined at local, regional, national, and international levels. The interconnections among institutions, and how these interconnections influence local resource use, are major topics for research.

Although researchers from many fields within the social sciences have been involved in development of political ecology as a framework for analysis, the major advances have taken place in geography and anthropology. Blaikie has continued to make important contributions (1989, 1995), and within geography the writings of Bebbington (1996, 2000), Bryant (1992, 1997), Peet and Watts (1996), Schroeder (1993, 1996), and Turner (1997) represent outstanding examples of political ecology research.

Anthropologists have been a bit more reticent at framing their research within political ecology, and Vayda and Walters have recently published a sharp critique of the field (1999). Nevertheless the early work of Sheridan (1988) and Stonich (1993) made convincing arguments for its use with anthropology. Escobar (1999) has used critical social theory to develop a "poststructuralist political ecology" to critique current ideas related to development and conservation.

Political ecologists will find that the influence of politics and history, as well as the scalar approach used in the case study, articulates well with their perspective on human-environmental relations. What is different here is the emphasis on ecology and the lack of an institutional framework to guide the analysis.

African Pastoralism: Competing Views and Alternative Futures

The way in which pastoral people utilize the resources in what are generally considered marginal lands has been an important topic of discussion in the literature on pastoral and nomadic peoples for over five decades. Central to this body of literature is the examination of mobility as a means of adapting to the patchiness of the resources and the spatial and temporal variability of the environment.

One can view the examination of mobility among livestock-keeping peoples as beginning with Evans-Pritchard's (1940) study of the Nuer in Africa and Lattimore's (1940) historical account of pastoralism in China and continuing up to the present (for recent accounts, see Dyson-Hudson and McCabe 1985; Bassett 1986; Ingold 1987; Grayzel 1990; Khazanov 1994; McCabe 1994, 2000). Although the theoretical thrust of the arguments has changed through time there are a number of issues that have dominated the literature. Among the most prominent are: (1) the causes of spatial mobility (the influence of ecological versus social and political factors dominates this discussion); (2) the impact of mobility on social organization; (3) the value of constructing topologies of movement patterns and the criteria used in constructing topologies; (4) the conceptual advantage of disarticulating exploitative strategies from spatial mobility (much of this literature involves comparisons of mobility among pastoral nomads and hunters and gatherers; for overviews see Johnson 1969; Salzman 1969, 1971, 2004; Irons and Dyson-Hudson 1972; N. Dyson-Hudson and R. Dyson-Hudson 1980; Fratkin 1991; Fratkin et al. 1994; Ingold 1980, 1987; Khazanov 1994).

Today, discussion and debate concerning mobility, land use, environmental degradation, and development center around two competing views of pastoral peoples, particularly the pastoral peoples of Africa. One view of African pastoralism is rooted in an understanding of rangelands as equilibrium-based ecosystems; here nomadic pastoralism is considered an anachronism, a way of life that must fade into the past for development to occur. Steven Sandford (1983) has called this perspective the "mainstream view" of African pastoralism, because it is so ubiquitous, has dominated the thinking about pastoral development (and still does), and underlies the assumptions upon which this thinking is based.

The alternative view is based on an understanding of arid and semiarid rangelands as nonequilibrium ecosystems; here nomadic pastoralism is seen as an efficient and productive adaptation to instability, and mobility is viewed as beneficial, if not critical. Adherents of this position advocate that past development efforts have failed because they were formulated on false assumptions about the structure and dynamics of the environments in which they were implemented. Development efforts need to take advantage of the "opportunistic" mobility patterns of the pastoralists, and institutions, from marketing to governance, must be built on an appreciation of instability of nonequilibrium-based ecosystems.

The Mainstream View

An understanding of equilibrial ecosystems is based on the assumption that "conditions outside the system of interest are relatively stable over time, allowing the internal processes of the system to play out or equilibrate, and regulate system structure and dynamics" (Ellis, Coughenour, and Swift 1993, 31). Coupled with the notion of equilibrium is the theory of plant succession proposed by Clements. Successional theory stipulates that one association or community of plant species will replace another in an orderly and directional process until a final or "climax" stage has been reached. In equilibrial systems homeostasis is maintained through biotic interactions.

Mature ecosystems can be returned to earlier successional stages through disturbances. A commonly cited example of this is cutting down a mature forest, then replacing the climax forest with crops or tree species associated with earlier successional stages; after abandonment the field will undergo a series of changes in vegetal associations until the climax stage is again achieved. In African rangelands the most commonly cited "disturbance" that returns a climax association to earlier successional stages is the overexploitation of vegetal associations by herbivores, almost always associated with domestic livestock. The "carrying capacity" of the rangeland is simply defined as the number of herbivores that can be supported by the vegetation for a period of time (usually a year) without degradation. In a "natural" grazing system the number of grazing animals will be controlled by the availability of forage and a stable state will be achieved; this is often referred to as the "ecological carrying capacity" (Behnke and Scoones 1993).

In rangelands grazed by livestock the ecological carrying capacity can be exceeded through the ability of humans to buffer the limitations of forage. This overexploitation will result in a change in the species composition of the vegetation to an earlier successional stage, in which less palatable grasses replace the more palatable ones. If this process continues, soil composition will be impacted, and bare soil will be exposed and subject to erosion. Eventually this "degradation" will lead to "desertification."

This understanding of range ecology was adopted by range managers and development planners throughout the semiarid and arid regions of the world. Range managers could maintain the health and productivity of the rangeland by adjusting the number of livestock to the state of the vegetation. When the process of degradation was observed the manager could remove some livestock from the range and the range would return to a more healthy climax or subclimax stage.

As Steven Sandford has observed, although this paradigm of range ecology was accepted by range managers throughout Africa, most of the livestock-keeping peoples who inhabited the African rangelands did not share this perspective on how to manage grazing lands (1983). Pastoral peoples living in Africa primarily managed their rangelands through communal grazing systems. Although access to grazing, water, and other resources in these traditional systems is always regulated, Western-trained range managers often viewed traditional management systems as inefficient, destructive, and sometimes "irrational," if they recognized them at all.

From early in the colonial period pastoral peoples were viewed as disruptive elements that contributed little to the economy of the region where they lived. The idea that East African pastoralists had an "irrational" attachment to their cattle was introduced by Melville Herskovits in an article published in 1926 entitled "The cattle complex in East Africa." Herskovits had never been to Africa, and the point he was trying to make was that the culture area approach in anthropology had applicability outside of North America and could be expanded to include values. Despite the fact that Herskovits was limited to secondary sources, principally missionaries' and explorers' reports, the perception of East African pastoralists as people who kept large herds of cattle primarily for prestige and who would not eat nor sell their cattle perpetuated the myth that traditional cattle herders in Africa behaved in an economically "irrational" manner.

Given the mainstream view of rangeland ecology held by most rangeland experts in colonial Africa the only way to save the rangeland resources was either to remove the pastoralists from them or to reduce the number of livestock grazing on these communal areas.

Destocking programs were common throughout the colonial period, and many pastoral peoples came to view state governments as little more than powerful raiders.

The mainstream view of rangeland ecology was given support by Garrett Hardin's "Tragedy of the commons" article published in 1968. Hardin used a commonly managed grazing system to illustrate his argument that it is in the economic best interest of the individual to overexploit a commonly managed resource, and he used a herder managing livestock in a communal grazing system to illustrate his point. He argued that it is in the best interest of a herder to accumulate as many livestock as possible without regard for range condition because the benefits of such a strategy accrue to the individual, but the costs of declining rangeland productivity are shared by the group. His argument was flawed, because as mentioned previously, traditional management strategies are not open access systems, but this lent further support to those who viewed pastoralists as inherently destructive to rangeland ecosystems.

The view that pastoralists will, by necessity, destroy the environments in which they live is classically illustrated in the writings of L. H. Brown and Hugh Lamprey. Brown was a former chief agricultural officer for Kenya and was a rangeland consultant to the Ford Foundation when he wrote "The biology of pastoral man as a factor in conservation" (1971). In this often cited article, Brown argues that the dependence on milk as the principal source of subsistence for many pastoralists requires that large herds be kept. A growing human population requires a growing livestock population, and it is impossible to maintain range conditions under these circumstances. Although Brown recognized that subsistence, rather than prestige, was the principal reason for the large livestock herds raised by pastoral people, the implications for range ecology were the same.

Lamprey was an extremely influential ecologist who has spent most of his life working in the East African rangelands. He has argued that wildlife lives in balance with the range resources, but that domestic stock managed by pastoralists will almost always exceed the carrying capacity of the rangeland and drive the systems into a degraded state that will eventually lead to desertification. He uses Brown's work to support his position, which can be succinctly stated as follows: "In balance it seems that the symbiosis of pastoral man and his domestic stock has been very successful as a survival strategy in the short term. In the long term it appears less successful since it tends to destroy its own habitat" (Lamprey 1983:656).

Lamprey felt that pastoral peoples commonly adjusted their management strategies to degraded conditions by (1) moving to areas not exploited by livestock for long periods of time, or (2) changing the mix of livestock herded to correspond to the forage resources available in the degraded environment (substituting camels and goats for cattle, for example).

The view of equilibrium-based ecosystems, successional theory, and Hardin's tragedy of the commons argument coalesced into the "mainstream view" or what Ellis and Swift have called the "pastoral paradigm" (1988).

The Alternative View

The alternative view is based on an understanding of arid and semi-arid rangelands as persistent but nonequilibrium-based ecosystems. This view is coming to be accepted by many ecologists working in the arid rangelands throughout the world and by many social scientists working with peoples living in these areas. It has important implications both for understanding pastoralist behavior and decision making and for development planning.

✱ In nonequilibrium ecosystems, livestock populations will remain well below their carrying capacity due to periodic droughts or disease outbreaks. The boom-and-bust cycle of livestock-based economies that has been observed by those working with pastoral peoples is to be understood as a normal and expected outcome of the system's unstable dynamics. In terms of development planning this means that efforts designed to limit livestock numbers, the core feature of almost all livestock development projects, will have little or no impact on the structure of the plant communities. In other words, they will not reverse what has been perceived as degradation; they will, however, result in significant impoverishment to the local people, the intended "beneficiaries" of the "development" projects.

If the "alternative view" is accepted, then opening up areas of rangeland that are underutilized because of lack of water becomes a very viable option to increase livestock numbers. The introduction of veterinary care should also be viewed with less trepidation than in the past (Scoones 1996). Pastoral movements are expected to be highly variable, with little or no regularity. Citing Westoby (Westoby, Walker, and Noy-Mier 1989), Scoones has said: "From one season to the next you cannot know what will happen. Contingent responses to uncertain events characterize pastoral strategies" (Scoones 1996:5). Despite the fact that pastoral strategies will be contingent responses to very uncertain conditions, two of the major advocates of the alternative view see pastoral land use as highly efficient and optimizing.

> The producer's strategy within non-equilibrium systems is to move livestock sequentially across a series of environments each of which reaches peak carrying capacity in a different time period. Mobile herds can then move from zone to zone, region to region, avoiding resource scarce periods and exploiting optimal periods in each area they use. (Benhke and Scoones 1993:14)

In the alternative view, development efforts should be designed to maintain mobility and flexibility while increasing animal numbers and offtake. National governments should try to strengthen pastoral institutions and retain some form of traditional tenure regimes. This is, indeed, an alternative view of African pastoralism.

But do African pastoralists behave in the way predicted by this new ecological thinking? Do their goals for the management of their herds and the future of their families correspond to the development agendas in the alternative view? To what extent do African pastoralists like the Turkana base their decisions about where and when to move, and how to manage their herds, on the ecology of the area in which they are living? To date there are few, if any, case studies that try to address these questions.

CHAPTER 3

Turkana

Environmental, Historical, and Social Overviews

According to the most recent Kenya census (1999), there are approximately 350,000 people living in Turkana District. This includes those living in the refugee camp in Kakuma, approximately 90,000 people, and people living in towns. Small groups of Turkana are also found in eastern Uganda, southern Sudan, and southern Ethiopia. Based on these figures I would estimate that there are approximately 300,000 Turkana people, and roughly two-thirds make their living primarily from pastoralism. The number of people who depend upon livestock as the basis for their livelihood varies over time and according to circumstances. Before the 1980s very few Turkana lived in towns or made their living from farming or fishing. The droughts and raids that swept through Turkana District in the early part of the 1980s created great hardship, and massive relief efforts were made to help prevent famine. By 1985 about half of the Turkana population was living in or close to famine relief camps. As drought stress lessened and relief camps began to close, many, but not all, of those who migrated into the camps returned to the pastoral sector.

During the 1990s drought and raiding again forced people into settlements, both for protection and for food. The situation remains dire today, but as conditions improve many who are now in settlements will rejoin those who have retained their pastoral livelihood. For pastoral people, the principal means by which people make their living is the raising of five species of livestock: camels, cattle, goats, sheep, and donkeys. The Turkana are some of the world's most mobile pastoral people with frequent shifts of homesteads and the subdivision of a family's herds into species-specific and production-specific units forming the core of their system of livestock management.

This chapter and the next present the ecological and social frameworks that will form the context for the analysis that follows. I want to first provide a broad overview of the climate, topography, and vegetation of the land where the Turkana people live; I also give an overview of Turkana history, social organization, and subsistence strategies. In the next chapter I will focus specifically on the area in which the South Turkana Ecosystem Project was based and where the families of Angorot, Lorimet, Atot, and Lopericho lived. Much of the ecological information presented is based on articles published by members of the STEP team, while background material on social organization and history draws on the work of Gulliver (1951), Lamphear (1992), and R. Dyson-Hudson and McCabe (1985) as well as on my own more recent research.

Overviews of Turkana history and social organization can be found elsewhere (Gulliver 1951, 1955; Dyson-Hudson and McCabe 1985), and they are a bit old-fashioned, reminiscent of ethnographic accounts of the 1970s and before. But they are important, and to encourage interested readers, I am including them here.

ENVIRONMENTAL BACKGROUND

The Turkana District of Kenya covers approximately 63,000 square kilometers (23,500 square miles), lying almost entirely within the Gregory Rift of the Great Rift Valley. It is bordered by Ethiopia in the northeast, by Sudan in the north, and by Uganda in the west. The eastern border bisects Lake Turkana, while most of the southern border is formed by the northern boundary of West Pokot District (see map 1.1, chap. 1).

Topography

The majority of the district consists of a low-lying plain with elevations ranging from 300 to 800 meters (1,000 to 2,600 feet). This plain is bounded to the west by the wall of the Rift Valley and on the south by the Samburu Hills. The plain is broken by a series of spectacular, isolated mountain ranges, which rise to elevations of 2,200 meters (7,000 feet), and lower lava hills and flows. Volcanic rocks cover approximately one-third of the district with the rest of the lowlands consisting of sandy and clay plains, lakebeds, and mountain foot slopes (Ecosystems Ltd. 1985).

The 4,000 foot contour has been identified as separating the better-watered highlands from the arid lowlands (Ecosystems Ltd. 1985). Using these criteria, 97 percent of Turkana District falls within the lowland category. These lowlands can be divided into three major drainage basins: the Lotikipi plains in the north, the Kalakol/Turkwell/Kerio lowlands in the center and south, and the Suguta Basin in the southeast.

There are at least fifteen mountain massifs rising above 4,000 feet, the largest of these being the Loima and Puch Prasir Plateaus located west of Lodwar. All these relatively less arid mountain ranges are extremely important to the pastoralists of the region as they provide critical forage resources during dry seasons and drought. Map 1.1 in chapter 1 shows the location of these massifs and ranges.

Climate

The climate is hot and dry. Analysis of precipitation records for a sixty-five-year period at Lodwar, the district capital (1923–88), revealed a mean annual rainfall of 180 millimeters (Ellis, Coughenour, and Swift 1993). The rainy season usually begins at the end of March and may last through June, with peaks of precipitation in April and May. A period of "short rains" may also occur during November. However, the amount, periodicity, and location of rainfall in Turkana District, as in all arid regions, is very difficult to predict. The annual rainfall record for Lodwar is presented in figure 3.1.

Further analysis revealed that the degree of variance in amount of rainfall from the long-term mean was striking, with a coefficient of variance of over 60 percent. In a different set of studies the STEP ecologists found that during the last fifty years rainfall has dropped 33 percent or more from the long-term average thirteen times, or once every three to four years (Ellis et al. 1987; Ellis and Swift 1988). Drought is often defined as a drop in precipitation to 85 percent of the long-term average (Stoddart et al. 1975), so the degree of drought

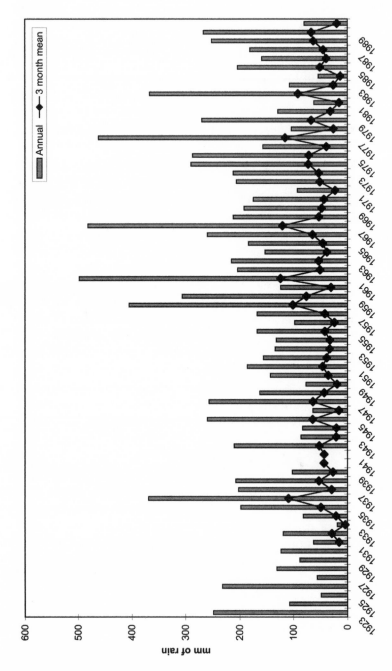

Fig. 3.1. Annual rainfall for Lodwar, Kenya, 1928–90, with three-month running mean

stress discussed here is severe. In addition, four of these drought sequences involved multiyear periods with rainfall below 33 percent of the long-term average. During a single year, drought plant biomass production was found to be less than half of that achieved in a normal year, and in a multiyear drought production may drop to one-third or one-quarter of normal years.

Temperatures and evapotranspiration rates are high. Data from Lodwar reveal a mean annual temperature of 29.3°C (84.7°F), with a mean annual maximum of 34.8°C (94.6°F), and a mean annual minimum of 23.7°C (74.6°F) (R. Dyson-Hudson and McCabe 1985). It should be noted that Lodwar is located in the center of the plains and is one of the more arid areas within Turkana District. The rainfall and temperature data should therefore be regarded as on the drier and hotter end of the rainfall and temperature curves of the district as a whole.

Vegetation

Although climate strongly influences vegetation dynamics, it is the interplay among climate, topography, and soils that determines the type of vegetation found in Turkana District. Rainfall increases with elevation while evaporation decreases, thus the higher elevations are significantly wetter than the low plains, and the amount and types of vegetation found in the highlands reflect these differences. Important exceptions to this relationship occur where groundwater is available and vegetation is not dependent upon precipitation for its water requirements.

Although over twenty years old, the vegetation classification published by Pratt and Gwynne in 1978 is still the most the common reference source for East Africa. They identified six ecoclimatic zones for Kenya and Uganda, and four of these occur in Turkana District. Ecoclimatic zones V (arid) and VI (very arid) comprise the vast majority of the land area in Turkana District. Ecoclimatic zone IV (semiarid) is found on the slopes of the Uganda escarpment and higher elevations in the western portion of the district and in the mountain range south of Lokichar, while small isolated pockets of ecoclimatic zone II (subhumid) are found on the tops of the highest mountain ranges along the Ugandan escarpment.

Pratt and Gwynne describe these zones as follows.

VI (very arid)

Rangeland of low potential, the vegetation being dwarf shrub grassland or shrub grassland with *Acacia reficiens* often confined to watercourses and depressions with barren land between. Perennial grasses are localized within predominantly annual grassland.

V (arid)

Land suited to agriculture only where fertile soils coincide with very favorable distribution of rainfall or receive run on; typically rangeland dominated by *Commiphora, Acacia,* and allied genera, mostly of shrubby habit. Perennial grasses . . . can dominate.

IV (semiarid)

Land of marginal agricultural potential . . . carrying as natural vegetation dry form of woodland and savanna, often an *Acacia-Themeda* association but including a *Brachystegia* woodland and equivalent deciduous or semievergreen bushland.

II (subhumid)

Forests and derived grasslands and bushlands, with or without natural glades

Water Resources

Although Turkana District is hot and dry, water can be found in springs and wells throughout the district. Lake Turkana is alkaline, and because there is no outflow, there is a continuous concentration of solutes. It also very high in fluorides, and although it would not be considered suitable for regular human consumption by international standards, a great number of Turkana people and their livestock use the lake as their regular source of drinking water.

There are no permanent rivers in the district, and only two rivers, the Kerio and the Turkwell, carry any surface flow for more than a few months each year. Most smaller drainages carry water only for a few days during the year when they are subject to flash floods during the rainy season.

Following heavy rains, water is often channeled into surface pools that can be an important resource for short periods of time. Pools are also found in rocky portions of stream and river courses. These pools can last quite a bit longer than the surface pools and provide important sources of water for livestock.

Natural springs are found throughout the district and are commonly saline, which adds to their value for watering livestock in need of salt, especially camels. Springs are more common in the southern part of the district than in the north. In southern Turkana very few areas are located more than 20 kilometers from a known spring (Ecosystems Ltd. 1985).

Wells dug through sand in riverbeds are a common feature of many drainages in Turkana District during the dry season and are a major source of water for people and livestock. A dry season survey by Ecosystems Ltd. recorded 23,640 wells in 1984; some of these were undoubtedly out of use at the time of the survey, but this figure does provide some indication of the abundance of wells in Turkana District. The widespread availability of water in the dry season is a key feature in understanding the pattern of land use and livestock management practiced by Turkana pastoralists.

Summary

Turkana District is hot, dry, and subject to periods of severe drought. Precipitation is unreliable and unpredictable in amount, duration, and periodicity. The vegetation is dominated by annual grasses, shrubs, and thorn-bearing trees. These features led one colonial officer to describe Turkana District as "the wildest and most worthless district in Kenya" (Rayne 1923, quoted in R. Dyson-Hudson and McCabe 1985).

The climatic conditions described here fit clearly within the framework of arid land ecosystems as nonequilibrium ecosystems. The aridity, frequency of drought, and extreme coefficients of variance in the precipitation pattern lead to unstable and unpredictable system dynamics. In the words of Jim Ellis: "Where droughts are frequent, population fluctuations prevent plants and herbivores from developing closely coupled interactions, ecosystem development and succession are abbreviated or non-existent, and ecosystems seldom reach a climatically determined equilibrium point. Uncertainty abounds" (1996:38).

THE PASTORAL WAY OF LIFE AMONG THE TURKANA

This section presents a very simplified and idealized picture of Turkana pastoralism, which may be helpful for those unfamiliar with pastoralism. Turkana history and a very detailed description of social organization (chaps. 6, 7), as it relates to the families, are discussed later.

The raising of animals lies at the heart of social and economic life for the Turkana people. Gulliver describes the importance of livestock to the Turkana as follows.

> They are the very stuff of life to all the people, involved in their labour, happiness, worry, and disasters. As a person grows up and develops and passes through the stages of individual life he is accompanied at every stage by stock—at first his father's, later his own. As soon as he is able a boy begins to herd his father's stock and to learn the traditional knowledge of husbandry. Girls learn to water, milk, skin, and cut up carcasses and cook the meat and work the skins. . . . The herds form a continuum in the development of the family. (1951:21)

Although written over fifty years ago, Gulliver's statement is directly relevant to the pastoral Turkana of today, including those discussed in this book. The Turkana keep five species of livestock. Cattle, *aite*, are primarily short-horned zebus, and while not very productive in terms of milk, they are better able to withstand drought conditions than most other cattle breeds. Camels, *ekal*, are a small variety of dromedary and are the most reliable producer of milk among the Turkana herds. Goats, *akine*, are of the East African Turkana type and are usually herded together with sheep. Sheep, *amesek*, are of the Somali black head and East African Maasai types. Goats are important for producing both milk and meat, while the sheep provide a critical supply of fat and meat, but not much milk. Donkeys, *esikeria*, are used as pack animals but are occasionally milked; only those people who are especially poor are reported to eat the meat of donkeys.

Each species of livestock has specific water and forage requirements. Cattle require herbaceous forage and should be watered every day; however, they are often brought to water every other day during the dry season. Camels require woody or leafy vegetation. Although they have a reputation for going long periods of time without water, they need to be watered every fourth or fifth day, if they are to provide a good supply of milk. Goats can eat both grazing and browsing forage and should be brought to water every other day. Sheep, like cattle, require grass and are herded and watered with the goats. Donkeys are also grazers but usually run freely and go to water on their own.

A Turkana herd manager must balance the needs of his livestock for forage, water, and protection with the needs of his family for food. Livestock and livestock-based products make up the bulk of the Turkana diet. Livestock are sold occasionally in order to purchase food, clothes, and other household necessities that are unusual for modern pastoral people, even in East Africa, where grain products provide most of the caloric intake for the human population.

During the rainy season, Turkana families typically aggregate, with their livestock, in large neighborhood associations called *adakar*. As the dry season progresses these associations break up, and families often separate their livestock by species and productive status. Milking camels and milking small stock usually remain with the herd owner and most of the people, while satellite herds of cattle, nonmilking camels, and nonmilking small stock pursue independent migration paths throughout the dry season.

A Brief History

The Turkana people emerged as a distinct ethnic group sometime during the early to middle decades of the nineteenth century. They form one part of a larger linguistic cluster commonly referred to as the Ateker group, which is composed of the Karimojong, Jie, and Dodos of Uganda, the Taposa and Jiye of southern Sudan, and the Nyangatom of Ethiopia (Gulliver 1951; Lamphear 1988, 1992; Muller 1989). All these peoples speak a mutually intelligible language and have a subsistence strategy based on pastoralism or agropastoralism.

History prior to 1500
Oral history and archaeological evidence suggests that the ancestors of the Ateker group lived somewhere in the southern Sudan prior to A.D. 1500 and most likely subsisted as hunting and gathering peoples. After beginning their southern migration these ancestral peoples incorporated both agricultural and pastoral pursuits, eventually separating into what Lamphear calls the "Agricultural Paranilotes," with a subsistence system based primarily on cultivation, and the "Proto-Koten-Magos" group, in which pastoralism was the dominant subsistence strategy (Lamphear 1976; Muller 1989).

History from 1500 to 1800

The period from 1500 through 1800 appears to have been character-ized by frequent splitting and fusing of ethnic groups and shifting alliances among the groups. Around this time the Karimojong estab-lished a distinct identity with a subsistence system based on the rais-ing of livestock, principally cattle, combined with small-scale agricul-ture. According to Gulliver, the people now called the Jie seceded from the Karimojong and moved to the vicinity of Koten Hill, while the parent Karimojong remained in the area around the Mogos Hills in Uganda (Gulliver 1951).

According to the oral histories collected by Lamphear, a group split off from the Jie and established themselves in the region near the headwaters of the Tarach River, in what is now Turkana District, dur-ing the Palajam generation set. This corresponds to the early part of the eighteenth century (Lamphear 1988). Their origin is recounted in the tale common to most Turkana. The myth relates the search for a lost bull by eight young men who traveled east from Jie country down the escarpment looking for the lost animal. When the young men reached the headwaters of the Tarach River they found the bull living with an old woman named Nayece who came from Karamoja looking for wild fruits. She welcomed the young men, built a fire for them and introduced them to the area. While there, they noticed the abundance of unoccupied grazing lands, and upon their return to Jie country they assembled other young men and girls and returned to the Tarach for the dry season. However, instead of returning during the wet season to their homeland, they decided to remain below the escarpment and eventually establish their own identity as Turkana, or "people of the caves." It is thought that this name was given to these immigrants because some of them lived in caves on Moru Anayese when they first settled there.

Lamphear points out that this origin myth is similar to others found in many other African communities but notes that it probably also recounts the actual movement of the people now known as Turkana down from Uganda and their Jie and Karimojong ancestors.

History from 1800 to 1880

During the remaining part of the eighteenth century the other Ateker groups expanded and occupied all of the grazing lands above the escarpment. By the beginning of the nineteenth century Turkana cattle camps began to push down the Tarach in search of new pastures upon which to graze their animals. As they moved westward the Turkana encountered other pastoral groups, some of which herded camels, livestock that the Turkana had never seen before.

These camel herders were the Cushitic-speaking Rendille and Boran pastoralists and lived in close association with the Maa-speaking cattle herders referred to as Kor (Samburu) by the Turkana. In addition, another group of cattle herders, the Siger, lived on Moru Assiger. Lamphear views these people as an example of the "old style" pastoral community described as a "heterogeneous, multilingual confederation" (1988:31). As the Turkana expanded eastward they began to both assimilate and disperse other groups. They first pushed to the north and east to Lake Turkana, and then to the south crossing the Turkwell River. It appears that by 1850 the Turkana occupied much of the territory they use today.

Although it is not clear exactly why the Turkana expansion was so successful, Lamphear feels that one major factor was that the Turkana herded short-horned zebu cattle rather than the humpless sanga, which the other groups, except the camel herders, depended on. The zebu is a hardier animal than the sanga, and more resistant to heat stress and disease. In addition, the lower forage and water requirements allowed for a more efficient exploitation of the arid rangelands of Turkana District.

Other factors contributing to the expansion may have been the superiority of Turkana weapons, which were forged by the Labwor and supplied by the Jie, and the emergence of powerful diviners (*emerons*) who were able to coordinate the recruitment of warriors from different sections into large armies (Lamphear 1992). The generation-set system also underwent alteration, but the impact of this change on the enhancement of the military is a matter of debate (see Lamphear 1992; Muller 1989).

In the 1870s the cattle held by people in many of the surrounding communities were stricken with a severe outbreak of contagious bovine pleuropneumonia, and as a result these communities were weakened considerably. This was followed ten years later by the rinderpest epizootic, which killed an estimated 90 percent of the cattle held by the Turkana's rivals, and shortly thereafter by an outbreak of smallpox. The dual catastrophes of livestock disease and smallpox resulted in social upheaval and allowed the Turkana, who had for the most part been spared from these diseases, to expand into the whole of what is now Turkana District.[1]

The Colonial Period to 1980

The first European to enter into the land of the Turkana was Count Samuel Teleki von Szek whose expedition reached Turkana in June 1888. He was preceded by Swahili caravans that first arrived in 1884, searching for ivory. About the same time that the Swahili arrived in the south of Turkana land, Ethiopian ivory hunters began arriving in the north. Within a few years this led to a period of conflict between the British and the Ethiopians over the colonial domination of the Turkana, which lasted until 1918.

Although Teleki did not have great difficulty with the Turkana, other explorers and adventurers did. Henry Cavandish was met with great hostility in 1897, and this began a series of conflicts between the British and Turkana. At this time the Turkana were administered by the British colonial administration of Uganda, which exerted little influence in the remote and hostile land of the Turkana. The Ethiopians, however, made frequent forays into Turkana land and considered the Turkana to be part of the Ethiopian empire under King Menelik. The Ethiopians traded with the Turkana and supplied them with rifles with which they fought the British and increased their raiding against the surrounding tribes.[2] By 1902, the British began to form an impression of the Turkana as "an incorrigibly aggressive people whose inexorable conquest would swallow up much of East Africa if not quickly checked" (Lamphear 1992:69).

The British responded by increasing the number and strength of the military patrols whose goals were both to subdue the Turkana and to prevent the Ethiopians from establishing a permanent presence in Turkana land. In 1905 a hut tax was introduced in the southern part of Turkana land. At this time the British also institutionalized the role of chief in an attempt to provide a structure with which to rule this "chief-less" society.

Following a series of epizootics that swept through Turkana land in 1908, the Turkana greatly increased their raiding activity against the neighboring Pokot. In 1910, the British responded by what was to become the common method of punishment for the recalcitrant Turkana: massive confiscation of livestock and the killing of those people who resisted. This strategy was so successful in the south that by 1910, the southern Turkana were considered "pacified," and the first civil administration was introduced. The civil administration lasted until 1918, when increased military activities by the Turkana forced the British to reinstate a military administration.

Although the British felt that the raiding problem was under control, this was also the time during which perhaps the greatest Turkana war leader of all time, Ebei, began to amass large armies and later directed raids against the Turkana's southern neighbors. The British launched a number of patrols with the expressed purpose of arresting Ebei and the emerons who were helping him. By 1917, Turkana armies had caused such disruption among the pastoral tribes to the south that the Pokot began to encroach onto the fertile highlands where British settlers had established large ranches and farms. Jeopardizing the security of the white settlers was unacceptable, and in 1918 a combined force of over 5,000 well-armed men, consisting of Sudanese troops, troops of the King's African Rifles, and levies composed of warriors from groups antagonistic to the Turkana, launched what came to be known as the Labur Patrol.

The Turkana fled before this massive army, occasionally fighting but all the time fleeing north. The Turkana lost hundreds of thousands of livestock, and hundreds, perhaps thousands, of people were killed or died of starvation or disease. Lamphear refers to the impact of this military expedition on the Turkana as "the most cataclysmic event this society had ever experienced" (1992:196). In addition to the losses suffered by the Turkana, the last remaining Ethiopian force was defeated, marking the end of Ethiopian involvement in Turkana District.

Although some raiding and counterraiding continued into the following decade, the Labur Patrol broke the military might of the Turkana. For the next decade the Turkana became the victims of increased raiding by the neighboring tribes, especially the Dassanetch of southern Ethiopia. In 1924 Ebei was killed in a Dassanetch attack along with the last remaining warrior segment of the armies assembled during Ebei's ascendancy. This marked the end of large-scale raiding and consolidated the power of the British colonial government.

In 1926, civil administration was reintroduced, and in 1928 taxes were reinstated. The period from 1929 until World War II appears to have been peaceful. The extent to which the Turkana were incorporated into the colonial administration is demonstrated by the fact that during the war the British supplied some of the Turkana with rifles, and companies of Turkana irregulars participated in the British invasion of Ethiopia.

Following the end of the war, the Turkana were disarmed by the British, but the Dassanetch retained their weapons. By the late 1950s the Dassanetch were launching frequent attacks against the northern Turkana. The inability or unwillingness of the British to protect the Turkana from Dassanetch raids rekindled a feeling of ill will directed at the colonial government. The increasing resistance of the Turkana toward colonial administration resulted in Turkana being classified as a "closed district," and the Turkana remained isolated from the rest of Kenyan society.

By 1960, only two primary schools and one mission had been established. By the early 1960s the Turkana had again taken their protection into their own hands and had resumed raiding, especially in the northern sections. In 1962, the King's African Rifles mounted another punitive expedition reminiscent of the 1918 Labur Patrol. Colonial administration ended with Kenyan independence in 1963.

During the period from independence until the 1980s, the development of Turkana District increased dramatically. Roads were built, missions established, and towns and trading centers grew. Irrigation schemes were established along the Turkwell and Kerio rivers, and the development of a fishing industry along the western shore of Lake Turkana was encouraged. However, this was also a period of unrest in the surrounding countries, which greatly facilitated the access to arms, and this period was characterized by intense raiding between the Turkana and their neighbors. This raiding continued into the early 1980s when my field research began.

SOCIAL ORGANIZATION: SOCIAL RELATIONS AND ACCESS TO RESOURCES

The Turkana, like many pastoral populations in eastern Africa, had no traditional form of political hierarchy based on chiefs and subchiefs.[3] Political influence was and still is gained through age, wealth, wisdom, and oratory skills. Although the traditional system influences life on a day-to-day level, the political structure of the Kenyan government defines administrative units and the relationships among them.

Turkana social organization can be seen as two systems of social relationships operating simultaneously. One system is based on social relationships, territory, and rights in pasture and water; the other is based on kinship, relationships among individuals, and rights in livestock and labor.

Social and Territorial Organization

The Awi

The basic unit of Turkana social organization is the *awi* consisting of a man, his wives, their children, and often a number of dependent women. These dependent women are usually close relatives who either have lost their husbands or have become so poor that their husbands could not support them. The human population is linked to the livestock population, and the awi is composed of these two interdependent populations. Livestock holdings consist of separate herds of camels, cattle, donkeys, and small stock (goats and sheep are herded together). Most awis will have herds and flocks of each of these livestock species, although the species composition of the herds and flocks will vary according to location, wealth of the herd owner, number and types of animals inherited, and the individual preference for certain species by the herd owner.

Each herd owner has a home area or *ere* to which he often returns in the wet season. A herd owner does not own rights to grazing in the ere, although he may own wells there. Most herd owners live and travel with two to five other herd owners and their families, forming what is referred to as a large awi or *awi apolon*. The composition of this unit changes frequently as individuals and families leave to join other awis or others come to join the awi apolon. Within this unit, herding responsibilities are often shared, and occasionally food is distributed among the awi apolon members. Sometimes herd owners remain together for years, and at other times they may share the large awi for only a few weeks.

During the wet season, awis congregate into temporary associations referred to as *adakars*. During good years these associations may consist of up to two to three thousand people and their livestock, and they may all live within a ten- to fifteen-square-kilometer area. People within the adakars remain together, for companionship and protection, for as long as the forage and water resources permit; this is often two to four months following the onset of the rains. In 1995–96 a new form of social organization emerged called *arum-rum*. This consisted of a large encampment of many herd owners under the leadership of a single man. People lived with a series of concentrically built thorn fences, and the young men within the arum rum were heavily armed. The whole encampment moved together as a unit, and I was told this was a response to the especially heavy attacks from the Pokot that occurred in the early and mid-1990s. I will discuss this new form of social organization in more detail in later chapters.

The Territorial Section

Each Turkana herd owner is a member of a territorial section. Membership confers equal access to the forage contained within sectional boundaries to all herd owners. Water sources, however, have many rules governing access, and these will be discussed later. The sizes of the sections vary considerably, from the relatively small territories of the Ngoboceros and Ngikebotok to the vast Ngikwatela range. There is some disagreement among scholars concerning the number of territorial sections into which Turkana land is divided: R. Dyson-Hudson and McCabe (1985), Gulliver (1951), and Lamphear (1992) state that there are eighteen or nineteen sections, while Muller (1989) feels that there are only fourteen. Our own work among the Turkana supports the sectional numbers and locations recorded in R. Dyson-Hudson and McCabe, and we will use that classification throughout. Map 3.1 shows the location of the named Turkana sections.

There is also disagreement among Turkana scholars concerning the extent to which the section acts as a grazing reserve. Gulliver and Lamphear state that the sectional boundaries mean very little, and individuals can bring their livestock into any section without concern for membership. I have previously written that Ngisonyoka boundaries are enforced (McCabe 1985, 1987; R. Dyson-Hudson and McCabe 1985). Because I had the opportunity to work among a number of sections as part of this study, I am able to confirm that both views are correct. For the Ngisonyoka, there are a number of conditions that must be met before a herd owner from another section is allowed access to the forage resources contained within the Ngisonyoka section. For the northern sections of the Turkana, the sec-

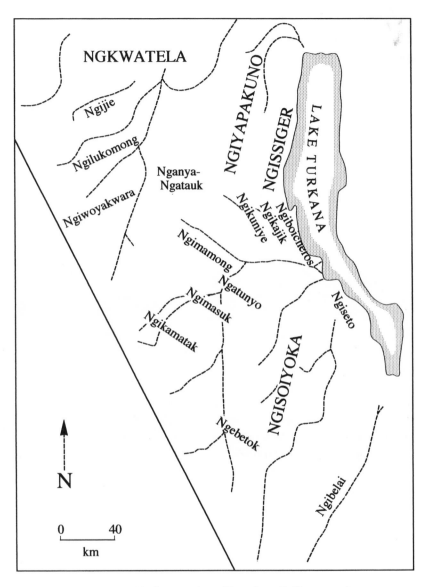

NGKWATELA

Ngijie

Ngilukomong

Ngiwoyakwara

Nganya-
Ngatauk

NGIYAPAKUNO

NGISSIGER

LAKE TURKANA

Ngikuniye

Ngikajik

Ngiboicheros

Ngimamong

Ngatunyo

Ngimasuk

Ngikamatak

NGISOIYOKA

Ngiseto

Ngebetok

Ngibelai

N

0   40

km

Map 3.1.   Turkana sections. (Data from Gulliver 1951.)

tional boundaries appear to have little meaning, with individual herd owners crossing through a number of sections in search of forage and water during the dry season.

One explanation for these differences lies in the proximity of plains and mountains in the south, which provide a variety of microhabitats that can be exploited throughout the year. Ngisonyoka have an agreement with another Turkana section that controls the Loriu Plateau, the Ngisetu, so that they can bring their cattle onto the plateau each dry season. Thus there is little need to engage in long migrations outside of their own territory. This contrasts to the situation in the north where vast distances exist between plains and mountains and it may be necessary to seek access to highland resources in different places each year depending upon where rain has fallen and the current security situation.

The Turkana Tribe

The largest unit of social organization among the Turkana is the tribe. Although the classification of ethnic groups as "tribes" has been strongly and appropriately criticized (Fried 1975), the notion of a single people, living in a contiguous area, among whom rules concerning raiding and violence are adhered to, and who consider themselves a single people, *Ngiturkan*, is a reality among the Turkana. It appears that in the past there were a number of ceremonial occasions in which tribal identity was reinforced (Lamphear 1992). Today tribal identity is reified by rules of appropriate behavior concerning raiding and banditry among the sections. The Turkana say that those who are not Ngiturkan are *emoit*, strangers and enemies; and enemies should be killed. This may be generally true, but it is certainly not always the rule. I will describe later how the Ngikamatak section of the Turkana have formed frequent alliances with the Mathenko section of the Karamojong; and during peaceful periods of time individual Ngisonyoka have developed personal friendships with individuals from the Pokot group.

Nonterritorial Aspects of Social Organization

Systems of kinship and stock associations form the basis by which individuals gain access to livestock and labor in times of stress. Every Turkana is born into the clan of his or her father, but women join the clan of their husbands upon marriage. There are twenty-nine clans among the Turkana, some of which are found throughout Turkana district while others are restricted within certain northern or southern sections (Gulliver 1951). Clans are exogamous, and membership is symbolized by the characteristic brand that appears on all camels and cattle of clan members. Clan membership provides an opportunity to form individual relationships of mutual assistance, but it does not obligate clan members to help one another. When I asked individual Turkana herd owners to whom they would look for help if they were in an unfamiliar place in Turkana land, the common response was that they would look for animals who had the same mark and then follow those livestock home, the assumption being that as relatives, the visitors would be provided with food and shelter. Although members of the same clan may be willing to offer hospitality, in time of real need it is the relationship between individuals that is of utmost importance.

There is neither a hierarchy of clans among the Turkana nor any corporate functions. The only clan assigned a separate status is the one from which the emerons (diviners) emerge. Although Gulliver states that anyone can become a diviner, our research indicates that all-powerful diviners come from a single clan.

Marriage provides a herd owner with a community of in-laws with whom he can form strong friendships. In-laws often form the core community in the awi apolon, and in-laws also play a critical role in the formation of the network of social relations, so important in time of stress (discussed later).

Stock Associations

Each Turkana herd owner is the center of a network of personal relationships based on the exchange of livestock. It is to individuals in this network that a herd owner turns in times of need. From the perspective of an individual herd owner, most members of this association will be agnatic or affinal kinsmen, but there may also be many nonrelated friends in this network. During peaceful periods individual Turkana will even form stock associations with members of other tribes.

Herd owners often keep some of their livestock holdings in the herds of their stock associates, thus reducing their risk in the event of a localized disaster, such as a raid. When food is needed for special occasions, herd owners frequently beg or borrow livestock from their stock associates, and a goat or a sheep will almost always be slaughtered when a stock associate comes to visit. Stock associates contribute to each other's bridewealth payments, and in the event of a major livestock loss a herd owner can expect his stock associates to help rebuild his herd.

One cannot overemphasize the importance of this institution among the Turkana. The long-term survival of a herd owner and his family may depend on these relationships, which are established over a lifetime. Other institutions, such as marriage, clan membership, or age and generation sets, provide the setting within which individuals form relationships, but they do little else with respect to securing access to livestock or labor. Livestock exchanges that may appear to make little economic sense to outsiders can be understood once one realizes that these individual relationships are reinforced through the transfer of animals.

Age and Generation Sets
Unlike many other East African pastoral peoples, the Turkana do not have a formal system of age grades through which males pass as they become older, nor is circumcision practiced for either males or females. Historical accounts indicate that an age-grade system was important to the Turkana in the past, and it is true that other Ateker groups do have a formalized age grade system (Gulliver 1951; Muller 1989). Men are initiated into adulthood during the *asapan* ceremony, and this ceremony appears to have played a central role in the age grade system. All men are supposed to go through the asapan before they are married, but today the ceremony is often held years after marriage takes place. During the ceremony a young man is given adult status by discarding his clothes and other material objects associated with his youth. These items are replaced by his "asapan father," a man chosen to help the young man become a responsible adult member of the community. Although the initiation ceremony seems to be of less importance than in the past, the relationship of the young man to his asapan father remains important.

The Turkana do participate in a system of alternating generation sets, composed of two groups: the stones or mountains (*emoru*) and the leopards (*eris*). All of a man's sons will be of the other set than his. In other words, if a man is of the stone generation set, then all his sons will be of the leopard generation set. In theory, there should be equal numbers of stones and leopards of all ages in the Turkana population. Stones and leopards will sit together as units during gatherings and at meat feasts. This division is one way in which people can be organized for specific tasks. During my research in Turkana the only time we saw this division utilized was during a period of intense bandit activity. The leopards remained in the plains to protect the awis, while the stones went to the mountains to find out where the bandits were to strike next.

Rights in Land, Water, and Livestock
The natural resources used by the Turkana are controlled through social relationships and social organization. Land has already been discussed. All Turkana have equal access to open sources of water, such as rivers when flowing, pools that form as rivers dry up, and springs. Shallow wells are also open-access water sources. Wells, on the other hand, are owned by the individual who dug the well; and only the well owner and his close kinsmen and friends to whom he has allowed use of the well may draw water from it. Wells may be located in dry riverbeds where water can be obtained by digging through the sand. They may also be placed where water can be obtained only after digging through sand, clay, and sometimes rock. Most wells are in dry beds of rivers or streams and usually fill in with sand when the river or stream flows. Thus each year the sand must be removed before they can be used.

The ownership of wells exerts a strong influence on land use patterns in particular locations. During the wet season, when access to water is not problematic, most areas can be used. But as the dry season progresses, access to forage may be limited to the livestock of those families who own wells in the area. Others have to move to locations where there is open access to water or where they own wells. Thus while forage resources may be available to all members of a section, in reality access to that forage is regulated for a large part of the year by controlling access to water.

SUMMARY

This completes the general overview, although I will expand on certain aspects of Turkana history in chapter 5 when I discuss the importance of understanding how raiding and violence have helped shape how Turkana people view the government and their neighbors. I will also return to the ecology of Turkana District (chap. 10) when I examine the movement of people and livestock in groups and differences among Turkana sections.

The context is now set for the more detailed accounts of Ngisonyoka ecology and individual decision making that follow. Large-scale ecological and political processes certainly impact how individual Turkana people live their life and make the decisions that they do. However, it is the more mundane, subtle changes that occur on a daily basis that are often missing from accounts in ecological anthropology and pastoralism. It is to this more detailed level of resolution that I now turn.

CHAPTER 4

Ngisonyoka Ecology

Land and Livestock

The Ngisonyoka are one of eighteen sections of the Turkana and occupy approximately 9,600 square kilometers[1] in the southwestern part of Turkana district. The ecologists of the South Turkana Ecosystem Project decided on the territory of the Ngisonyoka to define the ecosystem boundaries of their study. Thus from the very beginning of the project the ecosystem boundaries were defined by how the people living there conceptualized and used the land and its natural resources.

Ecologists, at least those with whom I have worked for over twenty years, seem less concerned with the issue of boundaries than do many anthropologists and other social scientists who comment on and critique the use of ecosystems. Although a tightly bounded ecosystem has some conceptual advantages, everyone recognizes that no natural ecosystem, however small, is self-contained. The general objective is to understand the ecological relationships that influence and to some extent determine the structure and dynamics of the ecosystem, not to precisely define its boundaries. The way resident people have conceptualized and use the land, and how access to natural resources is mediated through the system of social relations, are critical components to the overall study and are a major thrust of my research.

The relevant features of topography, climate, and vegetation of Ngisonyoka territory are presented here (as was done for the district in the last chapter). These features come together to form a partial basis for how the Ngisonyoka people conceptualize the landscape. In addition to the natural features, the Ngisonyoka-conceptualized landscape includes the quality and abundance of resources that their livestock depend upon, as well as an area's potential risks, especially disease and raiding.

The Environment

Topography

The topography of Ngisonyoka territory is highly diverse, consisting of plains, eroded lava hills and plateaus, and a chain of basement complex mountains that bisects Ngisonyoka from south to north. The two largest rivers in Turkana District roughly form the eastern and western boundaries of Ngisonyoka territory.

The elevation of the Rift Valley floor gently slopes from south to north; the base elevation of the southern plains is approximately 1,000 meters, while the elevation of plains surrounding the confluence of the Kerio River and Lake Turkana is approximately 700 meters. Elevations in the central mountain range vary from 1,000 to 1,400 meters, except for the two southernmost peaks, Loretit and Lateruk, which rise to approximately 2,000 meters. Although not as high as either of these two mountains, the massif called Kailongkol dominates the central portion of the chain. The northern end of the central mountain range consists of isolated highly eroded peaks that stretch from the town of Lokichar north to the district capital of Lodwar.

The material underlying this part of the Rift Valley floor consists of gneisses and schists, but the only area where this Precambrian basement is exposed is in the central mountains. The foot slopes of the mountains are made up of coarse sands, gravels, and boulders. Lava deposits laid down in the Miocene cover most of the eastern third of Ngisonyoka territory, and isolated volcanic cones are found frequently between the central mountain range and the Kerio River. Although technically not part of Ngisonyoka territory, the volcanic Loriu Plateau rises to the east of the Kerio River and is an important dry season area for Ngisonyoka cattle.

On the eastern side of the central mountains a series of coalesced alluvial fans made up of loamy sands forms a bajada that ends at the edge of the lava deposits (Coughenour and Ellis 1993). A dominating feature of the eastern plains is the frequency with which they are cut by dry river courses (*aiyanai*), which drain the central mountains during the rains. Water flows eastward to the Kalapatta River and then on into the Kerio, eventually emptying into Lake Turkana. The plains on the western side of the central mountains are slightly higher than the eastern plains, and the drainages are fewer but more deeply incised.

Climate

The climate of Ngisonyoka territory, like most of Turkana District, is hot and dry. Rain most often falls during the months of April and May, and sometimes a short rainy period occurs in November as well. Like in the rest of the district, the timing and duration of the rains is highly unreliable, and drought episodes are frequent.

The precipitation pattern in Ngisonyoka is characterized by a strong southwest to northeast gradient, with the southwestern portions of the territory receiving an average of 600 millimeters of rainfall per year, while the northeast receives approximately 150 to 200 millimeters. Higher elevations receive more precipitation than lower areas. In 1993 Michael Coughenour modeled the precipitation pattern in Ngisonyoka, taking into account both the rainfall gradient and the effect of elevation on precipitation (map 4.1).

Temperatures in Ngisonyoka are high, averaging between 29 and 30°C on an annual basis, with high rates of annual potential evapotranspiration (Coughenour and Ellis 1993). The vast majority of Ngisonyoka territory thus falls within the arid to semiarid zones as described by Pratt and Gwynn (1977). The semiarid zones consist of the southwestern portion of the territory and those elevations above 1,200 meters; the rest of the territory is classified as arid to very arid.

Vegetation

The vegetation found in the Ngisonyoka ecosystem is primarily determined by topography, geomorphology, and precipitation. In the more arid parts of Ngisonyoka territory it has been classified as dwarf shrub grassland, while the vegetation community in the arid zone has been classified as dry thornbush land (Coughenour 1992). Annual

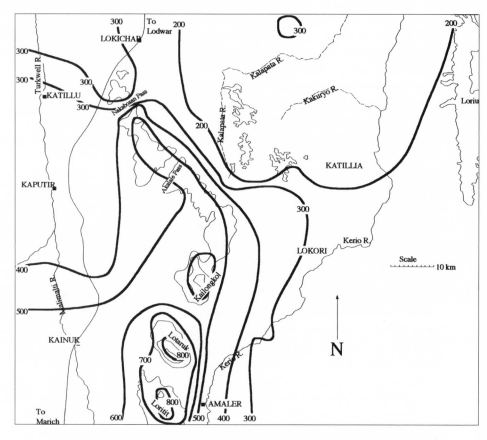

*Map 4.1.   Topography and precipitation of Ngisonyoka territory.
(Data from Coughenour 1993.)*

grasses predominate in areas that receive less than 350 millimeters of precipitation per year, while perennials are increasingly common in the higher and wetter areas. Groundwater woodland communities consisting primarily of *Acacia* species are found along the major river courses and form gallery forests along the banks of the Kerio and Turkwell Rivers. Large trees and an abundance of perennial grasses are found in the mountains.[2]

Increases in herbaceous and woody plant biomass roughly correspond to the increased precipitation gradient running from south to north in Ngisonyoka territory. Coughenour and his colleagues found that once an area has received 28 millimeters of rainfall each increase of 1 millimeter of rainfall would result in the production of 0.7 grams per square meter of herbaceous biomass (Coughenour, Coppock, and Ellis 1990). For Ngisonyoka territory as a whole, herbaceous production comprises approximately two-thirds of aboveground primary production. However, grasses and forbs only grow during the brief rainy season. As Coughenour pointed out, woody plants not only green up earlier than grasses and forbs, they also remain green for a much longer period of time. In terms of the energetics of the ecosystem, woody plants "serve to attenuate the short rainfall pulse into a more prolonged flow of energy to higher trophic levels" (Coughenour and Ellis 1993:385).

Unlike many pastoral systems in eastern Africa, burning does not exert a major influence on the structure of the vegetal communities found in Ngisonyoka territory. Ngisonyoka pastoralists, however, do burn those areas that will carry a burn, such as the mountain slopes and grasslands on the western side of the central mountains. Burning both reduces the amount of ticks living in an area and increases the amount of herbaceous vegetation. However, it has been estimated that only about 5 percent of Ngisonyoka territory is burned annually, and it only exerts a strong influence on vegetation at the local level.

NGISONYOKA CONCEPTUALIZATIONS OF THE LANDSCAPE

The Ngisonyoka see their territory as being divided into a number of regions (mostly named), based on landscape characteristics and vegetal communities; of course for the Ngisonyoka the importance of these regions relates to the kind of livestock that can be herded in these areas and types of management strategies used there. The identification of units described and how they are used is based on my research; but the description of the vegetation communities is based on the work of Coppock (1985). These regions are shown in map 4.2.

The Central Mountains

With the exception of the gallery forest found along the Turkwell River, the central mountains are the most productive region in Ngisonyoka territory. The vegetation is classified as bushed or

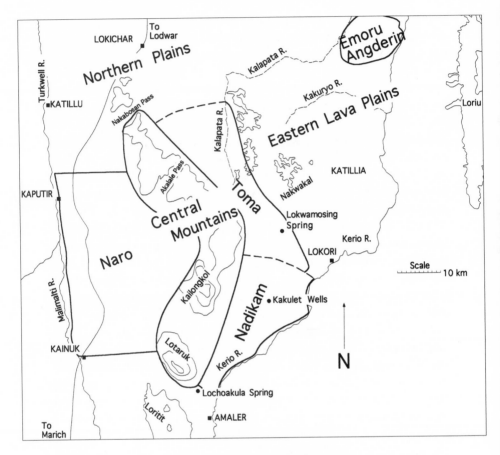

Map 4.2.   Ngisonyoka landscapes

wooded grassland, dominated by perennials. Grasslands are also
found on the lower slopes of the mountain range, but here perennials
are mixed with annuals. The common trees found in the higher elevations include *Euphoria candelabra*, *Boswellia* species, and *Acacia* species;
Acacia and *Commiphora* species are the common tree species growing
on the lower slopes.

The vegetation growing on the central mountains provide a critical
dry season resource for Ngisonyoka livestock, especially cattle. The
lower slopes are utilized by all livestock species, but only cattle move
up to the higher elevations. The principal reason for this is that

although the vegetation is abundant there, it is far from water, and people and cattle may have to walk ten to twelve hours each way to get water in the Kerio River or at the spring at Lochoakula.

Naro and the Western Plains

The Naro is the plain lying to the west of the central mountains. It extends approximately thirty-five kilometers north from the southern boundary of Ngisonyoka territory, to the confluence of the Korinyang and Turkwell Rivers. The Naro is the wettest of the plains found in Ngisonyoka and thus is quite productive. Unfortunately, much of the vegetation is in the form of a dense thornbush land dominated by *Acacia reficiens, Acacia mellifera,* and *Acacia senegal.* Abundant herbaceous vegetation grows under the trees and can provide good forage for cattle. One area of the thorn thicket has been kept clear of trees by burning and is an important site for Ngisonyoka grazing. This area is referred to as Kadengoi, and after the rains begin Ngisonyoka cattle often congregate there waiting for the grasses to grow in the Toma. Both disease and raiding are major constraints in utilizing the lush vegetation of the Naro. Ticks and tsetse flies (the vector for trypanosomiasis) are abundant here, and in addition to the proximity of the boundaries with both the Pokot and the Karimojong, it can be a very dangerous place for Turkana. In many respects the Naro is used as a drought reserve, although many people have told me that in times of peace it is much more heavily used than during the 1980s and 1990s. It is also quite possible that during a time of more intense use, the bush cover along with ticks and flies diminishes, and then there is less disease.

The northern section of the western plains, extending north from the Korinyang River to the town of Lokichar, does not have a specific Turkana name and is not extensively used. It is significantly drier and thus less productive than the Naro.

Nadikam

The area referred to as Nadikam by the Ngisonyoka lies to the east of the central mountains and extends north from the southern Ngisonyoka boundary to approximately the northern slopes of Kailongol mountain. Topographically it is an area of dissected foot slopes and volcanic cones. Precipitation is moderate in Nadikam with the southern portion of the region wetter and more productive than the north. Although grasses are found in sites throughout Nadikam, the area is known for its abundant woody vegetation, with *Acacia reficiens* dominating. Although the woody vegetation is dense in Nadikam it does not grow into the types of impenetrable thickets found in the Naro.

Nadikam is the traditional dry season area used by the Ngisonyoka and is heavily used by goats and camels. There are two major springs available for livestock in Nadikam, Lochoakula in the south and Lokwamosing in the north. There are also large wells at Kakulet, and water is generally available from the Kerio River (when flowing or later in pools). Like the Naro, the southern extension of Nadikam can be considered a drought reserve. People and livestock move into this region in bad dry seasons and certainly in times of drought. But like the Naro, it is dangerous. There was at one time a village near the spring at Lochoakula, complete with shops, a primary school, and a dispensary. Now they are all gone; only the deteriorating foundations can be seen barely emerging from the surrounding bush. The village was repeatedly raided by the Pokot until it was eventually abandoned, the iron roofing material on the school and dispensary carried away by the last raiding party.

Toma

The Toma is a flat sandy plain lying in the center of Ngisonyoka territory. The word *toma* means inside, and in this context it means the plain lying inside (or between) the central mountain range and the Kaweriweri plateau. Although relatively dry, there is a significant water subsidy that results from runoff from the central mountains during storms. The plains are cut by numerous streambeds each of which supports a small groundwater woodland and herbaceous community. The woodlands primarily consist of *Acacia tortilis, Acacia elatior,* and *Delonix elata,* while the herbaceous community is dominated by *Brachiaria, Erogrostis,* and *Dactyloctenium* species. The interdrainage areas called *atot* support a community made up of annual grasses and dwarf shrubs (*Indigofera* species).

The Toma is the principal wet season area of the Ngisonyoka families with whom I worked. The variety of forage species available here makes the Toma an ideal place to live during the wet season. There are few ticks and no tsetse in the Toma, and people will remain there as long as there is forage available for the livestock.

Eastern Lava Hills and Plains (Kakurio)

The entire eastern third of Ngisonyoka territory consists of volcanic deposits in the form of cones, flat-topped plateaus, and lava-covered plains. This is the most arid region in Ngisonyoka and thus is the region of lowest productivity. When it does rain, the Kakurio River drains the region, and the Ngisonyoka generally refer to this area merely as Kakurio. Annual grasses provide abundant forage during good wet seasons, but during dry seasons the lower elevations may be totally barren. The plateau regions here support small stands of *Acacia* and *Commiphora* species as well as small shrubs.

The area is best known for the abundance of salt springs found throughout this area. The Ngisonyoka rely on the springs to provide essential salts to their animals, especially camels; thus even though this area may only have sparse vegetal resources it remains an important area for livestock when herd owners feel their livestock require salt. People and livestock will also retreat to this area if raiding threatens the Toma. The Kaweriweri plateau protects it, and although the vegetation is sparse, it can support livestock for short periods of time.

The Loriu

The Loriu forms the northern section of an arc of high lava plateaus, rising to an elevation of 1,600 meters. Although the Loriu is not part of Ngisonyoka territory (it is within the boundaries of the Ngiesitou section), Ngisonyoka regularly use the Loriu for dry season grazing for their cattle. I have not visited the Loriu, nor have the STEP ecologists, due to its distance from the rest of Ngisonyoka area and the danger of raids (primarily from Turkana bandits, or Ngoroko). However, from low-level aerial reconnaissance it appears that large stands of annual and perennial grasses interspersed with trees and shrubs characterize the vegetal communities on the top of the plateau.

Water is a problem for people and livestock living on the Loriu, and the Ngisonyoka most often use it during the early part of the dry season. The area is also known for its abundance of wild animals, especially lions. I have been told on many occasions about cattle camps being attacked by lions, and if a herd boy is not both brave and vigilant, the chances of losing cattle to predators on the Loriu is great.

The Northern Plains

The region lying to the north of Lokichar and extending to the Ngisonyoka boundary near Lodwar is referred to here as the northern plains. There is no specific Turkana name for this area, although specific areas are named (e.g., Lochoa Ngikamatak, which means the well of the Ngikamatak, or Emoru Angderin, a small mountain used for dry season grazing). The region is much drier than the southern plains and mountains, and annuals dominate the herbaceous community. In the dry season there are large areas of mostly bare ground. Woody vegetation is dominated by *Acacia reficiens*, but the dwarf shrub *Indigofera* is also common. Near Lodwar, gravelly soils give way to fine sands that support annual grasses and can be quite productive during good rainy seasons.

The Rivers

Although not restricted to a particular region, large and small rivers and ephemeral watercourses crosscut the plains throughout Ngisonyoka territory. Although water may flow in these streambeds for only a few days a year (in some cases only a few hours per year), these watercourses are a very important component of the ecosystem. Most of the streambeds support stands of trees that provide essential dry season forage for goats and camels, while gallery forests grow along the banks of the large river systems.

In many of these river systems, water infiltrates into the sand and remains stored there throughout the dry season. Ngisonyoka access these pockets of water by digging wells, which allows for a much fuller exploitation of forage resources than would be expected in a desert environment (see next section).

RANGELAND PRODUCTIVITY

Before moving on to a description of water sources and livestock characteristics, it may be useful to bring the discussion of vegetation to a close by describing the average aboveground net primary productivity of the Ngisonyoka ecosystem. To a large extent the amount and quality of forage resources is a result of annual primary productivity of the vegetation. I return to this issue in both the analyses of individual herd owner mobility patterns and group movement, but here I want to summarize what is known from the ecological analysis.

Primary productivity of the vegetation in Ngisonyoka territory roughly corresponds to the north–south gradient in rainfall mentioned in previous sections. Ecologists working on the South Turkana Ecosystem Project made estimates of primary production for wet and dry season ranges for the early 1980s (Ellis et al. 1987), and Michael Coughenour refined the primary productivity data for the period 1982 through 1987 (1992).

The principal methods used for estimating primary productivity included ground survey, aerial survey, and the analysis of imagery gathered by remote sensing, in particular the advanced very high resolution radiometry (AVHRR), from which normalized difference vegetation indices (NDVI) can be derived.[3] NDVI is an indirect measure of plant biomass and is correlated with green leaf biomass. The combination of these methodologies allowed Coughenour and associates to estimate the spatial variability of primary productivity for all of Ngisonyoka territory.

The wet season range for most of the Ngisonyoka centers around the Toma, with approximately 18 percent covered in trees and about 70 percent classified as either bushed or wooded grassland. In a good rainfall year, woody foliage and dwarf shrub production averaged 100 kilograms per hectare, while herbaceous production was estimated at 1,800 kilograms per hectare.

The Ngisonyoka generally move to the south as the dry season intensifies, and the southern dry season range incorporates Nadikam, the mountains, and the Naro. Woody canopy covers about 40 percent of this area, and during a good rainfall year, woody foliage production is estimated at 200 kilograms per hectare. Herbaceous biomass consists of both annual and perennial grasses, and production here is estimated at 2,800 kilograms per hectare.

The Ngisonyoka also occasionally use the northeast lava plains during the dry season, especially following a good wet season. During good years herbaceous production, dominated by annual grasses, was estimated at 1,200 kilograms per hectare, while woody foliage production was estimated at 100 kilograms per hectare.

Because primary productivity closely corresponds to precipitation in arid areas, one can refer to the rainfall distribution map as an indicator of the spatial variability of primary productivity (see map 4.2).

WATER RESOURCES

Although Ngisonyoka territory is one of the more arid regions in Africa where the raising of livestock is the dominant way of making a living, access to water is generally not problematic. Surface water is available for only a few months of the year and in only a few locations, but numerous springs are found in the volcanic eastern third of Ngisonyoka territory. Further stretches of dry riverbeds where subsurface water accumulates and wells can be dug occur in many of the large and small drainages throughout Ngisonyoka territory.

The two major rivers that define the eastern and western boundaries of Ngisonyoka territory, the Turkwell and the Kerio, both arise in the highlands to the south of Turkana district. The headwaters of the Kerio River arise in the Cherangani hills and the Elgyeo escarpment, while those of the Turkwell River arise in the Uganda highlands to the south and west of Ngisonyoka territory. Both are areas of high rainfall, and thus these rivers are not dependent on local rainfall regimes. The Turkwell may flow for half the year, while the Kerio may flow for only a few months.[4] Before drying up completely water collects in pools and is accessible to livestock for many weeks and sometimes months. Water when flowing and in pools is available to everyone, and likewise there are no restrictions on who is allowed to dig shallow wells through the sand.

There are a number of smaller watercourses, which may only flow for a number of days during the year, but where subsurface water is available throughout the year. These rivers are referred to as *Angolol* (pl. *ngangololin*); examples of these systems are the Kalapatta and the Lokichar Rivers. Smaller river systems, referred to as *Ayonai* (pl. *ngayonai*), also contain areas where wells may be dug, but the water in these systems may only be available for a number of months following the end of the rains. Many of the watercourses that drain the central mountains fall into this category.

Anyone who wishes to can dig a shallow well through the sand in these river systems, but as the dry season progresses, the water-bearing stretches of the dry riverbeds, *Echor* (pl. *ngichorin*), become restricted, and well sites are owned by individual heads of households. Here wells may require digging twenty or thirty feet through sand, clay, and even rocks. The individual who digs the first well in these locations owns the well, and rights are passed down through the male line. Close relatives of the "owner" of the well also have access to the well, as do friends of the owner once permission is given. Fights may occur when unauthorized individuals attempt to use the well and "steal" water.

Springs, or *Ngichowai* (sing. *echowa*), are common in the eastern third of Ngisonyoka territory, although some are also found in the Central Mountains. Some springs flow throughout the year, while others last six months or less. The water flowing from many of the springs, especially those in the volcanic east, is strongly alkaline and supplies a critical source of salt for camels and goats. Anyone can use the springs when water is flowing, but because they frequently occur in the more arid parts of Ngisonyoka territory, access is restricted by the availability of forage. Other water sources are very small springs, *Ngidilite,* and areas where only a small amount of water will be available, *ngakujan;* water from these sources is commonly used only for human consumption and sometimes for very young small stock.

Although access to the forage resources in Ngisonyoka territory is open to all members of the section, access to water is not. Those without access to deep wells during the dry season must depend on the unrestricted sources of water. In effect, this limits the number of families who can use the Toma and parts of Nadikam during the dry season.

The only areas where there are serious constraints on water availability are in the Central Mountains and on the Loriu. Here, under drought conditions, cattle may have to walk for twelve hours from a grazing area to a water source. These areas are limited to cattle because camels cannot negotiate the rocky terrain, and the distances are too far for small stock to travel.

The natural resources of the Ngisonyoka are thus both socially constructed and regulated through the system of social relations. This is not surprising, but some ecological studies have failed to give proper emphasis to the important degree to which natural resources are embedded within the social fabric of everyday life. Ecological studies that ignore the social and cultural rules by which access to resources is gained provide little insight into how these resources are actually used. In the description of movements and decision making for the four herd owners described later in this book, the importance of access to wells will become obvious.

LIVESTOCK

Like most Turkana, Ngisonyoka pastoralists raise five species of live-stock: camels, cattle, sheep, goats, and donkeys. This represents one of the most diverse assemblages of livestock kept by pastoralists in Africa (Coppock 1985). Each species has different forage and water requirements, as well as different labor requirements. Each livestock species also has different fertility rates, is subject to different diseases, and produces different amounts of food. Although it is not unusual for a herd owner to specialize in one species, most Ngisonyoka families will have some livestock of all species in their herds. A herd owner may adjust the species composition of his herds to conform to specific needs of the family. For example, if a herd owner has lost a significant percentage of his herds due to drought, disease, or raiding he may exchange most of his remaining large stock for goats, which will reproduce much faster than camels or cattle and build up his live-stock holdings. Once he has reestablished himself, he may then exchange goats for large stock, which will not reproduce as fast but will supply a steady source of food (e.g., camels) and require far less labor (e.g., cattle).

Camels

The camels raised by the Ngisonyoka are small species of dromedary, *Camelus dromedarius*. They rely primarily on woody or leafy vegeta-tion for forage and require inputs of salt supplied by drinking water from the numerous salt springs in the eastern section of Ngisonyoka territory. Although camels are known for being able to go for long periods of time without drinking, this practice results in low milk pro-duction, and because camels are the most reliable suppliers of milk for the Ngisonyoka, care is taken that camels are brought to water every four to six days. While camels are the mainstay of the Turkana milk supply, they are slow to reproduce, dropping calves only every two to three years.

Cattle

The cattle raised by Ngisonyoka are a mixture of the Karimojong breed and small short-horned zebu breeds of *Bos indicus*. These types of cattle are relatively resistant to drought, but milk production is quite low. Ngisonyoka territory is, in general, not good country for cattle, but they do survive and during periods of high rainfall can put on weight rapidly and produce good yields of milk for a few months. Because they depend on grasses for the bulk of their diet, they must be moved from the plains to the mountains shortly after the dry season begins. Water can be a real problem for cattle as they need to be driven to water once every two days, and while in the mountains, this can result in very long treks to water sources. The mountains are also a dangerous place, both for the cattle and for those herding them. Predators are often abundant and raiders close by.

One of the questions that puzzled me during the entire first year of my research was why the Ngisonyoka kept cattle in the first place. They did not seem to contribute much to the daily subsistence of the people, and because they required grass to eat they were often in mountains with little accessibility and considerable risk. However, cattle clearly occupy a special place in the culture and social life of the Ngisonyoka. There are a number of ritualistic activities associated with cattle that are not associated with other animals. Each Turkana man has a "dance ox" that is specially decorated. Turkana men sing songs to their dance ox, praising its looks and strength, and call out the name of their dance ox when entering battle. Turkana men also have an "ox name" by which they are frequently known. A small bow and arrow are employed when bleeding cattle, similar to those used by other cattle-keeping people in East Africa, in contrast to the small cutting tools used to puncture veins in camels and goats when they are to be bled.

Some anthropologists have noted that cattle play a special role in the lives of Turkana women also. For example, Broch-Due mentions that cattle are also "cherished, because they illuminate various aspects of exemplary human fertility, nurturance, and care," and that conceptually human and cattle reproduction are "interwoven" (1999:69).

Cattle also play a unique role ecologically. They put on weight fast, and following the onset of the rains they can capitalize on the brief period of time when the landscape is bursting with green grass. None of the other livestock species can do this. This line of reasoning could potentially lead to the classic trap of ecological functionalism, but to deny the unique attributes of cattle physiology for a purely "cultural" explanation of their role in the life of the Ngisonyoka would be as egregious an error. It is the interplay of culture and ecology that explains why cattle are so valued, and why they occupy such a unique and to some extent enigmatic place in Ngisonyoka life.

Small Stock

The principal sheep breed raised by Turkana pastoralists is the black-headed Somali, although other breeds (mainly fat-tailed brown Maasai) are also raised. Sheep, like cattle, depend upon herbaceous forage and are thus more vulnerable to drought stress than browsing livestock. Sheep are used primarily for meat, barter, and sales. Sheep are also milked, but because they are often stressed nutritionally they are bled rarely, if ever. Most sheep are of the fat-tailed varieties, and their tails are composed almost exclusively of fat. The Turkana eat the tails when high inputs of calories are considered desirable, such as when someone is sick. Friends have told me that sheep are considered the "hospital" of the Turkana.

Goats (*Capra hirtus*), unlike sheep, are mixed feeders and are very well suited to the Turkana environment. The type of goat herded by Turkana pastoralists is referred to as "small East African," characterized by relatively small stature and small erect ears (Pratt and Gwynne 1977). Goats and sheep are generally herded together, but goats far outnumber sheep in these mixed flocks. Sometimes large castrated goats will be decorated like a dance ox and given special status within the flock. Goats are used for milk, meat, and blood, and are sold. Turkana consider goats a very important livestock species; they often form the bulk of a man's livestock herds.

Donkeys

The donkeys kept by Turkana are of the East Africa type (Pratt and Gwynne 1977) and are small, gray in color, with a distinctive black stripe that runs down the animal's back. Donkeys are known for their exceptional tolerance to dry conditions, with only camels being superior in withstanding periods of water deprivation. Primarily grazers, they can eat extremely coarse grasses that are unpalatable to other mammals, and they extract far more nutrients from standing dead grass than other livestock species.

Donkeys are not herded like other livestock, and Turkana generally allow donkeys to forage and go to water on their own. They are used almost exclusively as beasts of burden and are rounded up when a family moves. Occasionally donkeys are milked, but only children drink the milk. Corralling donkeys and trying to milk them is a difficult task, and few people make the effort. Donkey meat is generally considered unsuitable for human consumption, although poor people are known to eat it. Turkana say eating donkey meat imparts a particular smell to the body, and people who are known to consume donkey meat are often singled out for ridicule.

Livestock Characteristics and Feeding Ecology

Livestock form the basis for nearly all livelihood activities for the Ngisonyoka and are the medium through which energy and nutrients are transmitted from the environment to the people. Neville Dyson-Hudson once said that you could not understand a pastoral people without understanding the livestock and how they were managed. A detailed description and analysis of how these animals were managed follows in the case study of the individual herd owners. What they ate, how much and how frequently they drank, how far they walked, and how often they reproduced are summarized later. The basic characteristics of Ngisonyoka livestock described here are based on the work of Layne Coppock.

Layne Coppock had the arduous task of calculating the exact dietary intake of each of the species of livestock herded by the Ngisonyoka. In order to do this he accompanied different herds for seventy-five days during both wet and dry seasons. While he was with the herds, he counted how many bites of each vegetal species a particular animal ate during two-hour trials at different times of the day. In all he counted 350,000 bites; the results of these studies provide the best record available anywhere on the diet of livestock managed under traditional conditions. In addition to the physical difficulty that the work required, he also had to endure the questions and comments of Turkana herd boys, who found this activity amusing, to say the least.

Management decisions concerning separation of species-specific herds and the migratory patterns of these herds are often directly related to dietary requirements. Coppock's research demonstrated that herbaceous vegetations composed 96 percent of cattle diets; 67 percent of sheep diets; and 71 percent of donkey diets. Camels, in contrast, had a diet that consisted of 72 percent dwarf shrubs and 23 percent other woody vegetation. Goats ate a combination of herbaceous vegetation, dwarf shrubs, and other browsing vegetation (this data is included in table 4.3 at the end of the chapter).

Another aspect of dietary intake is the relationship of diet to food produced for humans. There is a fairly strong relationship between the amount of green vegetation eaten by livestock and their overall condition and milk production. In Ngisonyoka territory, woody and leafy vegetation (browse) remains green for much longer than does herbaceous vegetation. Dwarf shrubs remain nutritious well into the dry season, and the large trees that commonly grow along dry watercourses are able to tap into groundwater through their root systems. Thus, some trees may remain green throughout the year. Camels are large enough to be able to feed on the leaves of these trees, and their diet includes far more green vegetation than the diet of any of the other livestock species. It is not surprising that Ngisonyoka place such a high value on camels, or that camels form the basis of the Ngisonyoka milk supply.

The availability of forage, livestock diet, access to water, and distance traveled during the day all influence the health of the animals and the amount of food they can produce for the human population. Some basic characteristics of the livestock herded by the Ngisonyoka are summarized in table 4.1.

The watering of livestock is one of the most labor-intensive tasks undertaken by Turkana pastoralists. When all family members are together at the awi, it is usually the job of young women to lift water from wells to troughs from which the livestock drink. When the livestock are separated from the awi, young men perform this task. In general, small stock and cattle are watered from wells, while camels are driven to springs and sometimes to the Kerio or Turkwell Rivers. Camels need salt in their diet, and by taking them to springs to drink, they not only get access to salt but can drink freely, thus saving the labor that would need to be expended if they were drinking from wells. Table 4.2 summarizes data collected by Coppock on the seasonal watering regimes and intake of livestock herded by the Ngisonyoka.

TABLE 4.1. Characteristics of Livestock Herded by the Ngisonyoka and Seasonal Change

	Seasons			
	Wet	Early Dry	Mid Dry	Late Dry
Camels				
Live weight (kg)	399	401	410	404
Milk yield per day[a] (ml)	1,668	1,572	1,214	552
Distance traveled per day (km)	15	21	23	19
Cattle				
Live weight (kg)	184	190	210	194
Milk yield per day (ml)	1,133	662	775	279
Distance traveled per day (km)	11	31	15	16
Sheep				
Live weight (kg)	25	27	29	26
Milk yield per day (ml)	small amounts throughout the year			
Distance traveled per day (km)	12	16	12	15
Goats				
Live weight (kg)	24	27	30	26
Milk yield per day (ml)	177	75	128	63
Distance traveled per day (km)	12	16	12	15

Source: Data from Coppock et al. 1986a.

[a]This is the amount of milk available to humans—total milk production is twice this amount. These figures are based on Galvin 1985. Similar figures are found in McCabe 1984, but because Galvin's study of milk production spanned a longer period of time and is generally referred to in the ecological studies, I am using it here.

TABLE 4.2. Seasonal Water Intakes for Ngisonyoka Livestock

	Watering Frequency[a]	n	Liters Consumed	Intake ml/kg/Day
	Mid Dry Season			
Camels	4:1	5	96.4	47
Cattle	1:1	5	34	80
Sheep	1:1	5	5.6	97
Goats	1:1	5	5.5	88
Donkeys	1:1	5	34	87
	Wet Season			
Camels	5:1	5	116.8	48
Cattle	0:1	8	29.6	164
Sheep	2:1	9	11.4	153
Goats	2:1	10	12.1	168
Donkeys	1:1	6	29.7	79

Source: Data from Coppock, Ellis, and Swift 1988a.

[a]Ratio of feeding days to watering days.

TABLE 4.3.   Livestock Characteristics Related to Diet and Fertility

Characteristics	Camels	Cattle	Goats	Sheep	Donkeys
Dietary intake					
Herbaceous (%)	5	96	36	67	71
Dwarf shrub (%)	72	4	27	28	28
Other browse (%)	23	0	37	5	<1
Fertility					
Years to first calf	3–4	2½–3	1+	1	2½–3
Gestation period (months)	13	9	5	5	9
Years between births	2–3	1–2 (usually 2)	½–1	½–1 (usually 1)	1–2
Minimum doubling time (years)	9	6½	3	3	no data

Source: Data from Coppock et al. 1986a; Dyson-Hudson and McCabe 1985; and Dahl and Hjort 1976.

Livestock reproduction lies at the heart of the sustainability of the Ngisonyoka pastoral system. Differential reproductive rates for different livestock species allow the Ngisonyoka to manipulate herd growth and milk production, depending upon a family's needs at a particular time. Goats and sheep usually give birth once a year, although in very wet years they may actually drop kids or lambs twice within a single calendar year. They also reach reproductive age after only one year. Thus a flock of small stock can grow quickly, and it is, therefore, not surprising that Turkana consider these animals extremely important when trying to increase livestock numbers.

Camels produce milk steadily and can give milk even when conditions are extremely harsh. However, they reproduce slowly, only giving birth once every two to three years. They also take between three to four years to reach maturity. Cattle give birth once every one to two years and can reach reproductive maturity after two to three years. Data relating to livestock reproduction for Ngisonyoka livestock are summarized in table 4.3.

The feeding ecology of individual livestock species is an important component in decisions made by herd owners about how they manage their livestock. It figures critically in decisions about where to move, when to move, and whether to divide the herds or keep them together. This factor must be kept in mind to understand herders' movements and decision making (chap. 7).

CHAPTER 5

Cattle Bring Us to Our Enemies

> Nuer say that it is cattle that destroy people, for more people have died for the sake of a cow than for any other cause.
> —E. E. EVANS-PRITCHARD, *The Nuer*

My focus has been on the importance of ecology, especially arid land ecosystems as nonequilibrium systems, in understanding how the Turkana people use the land. Now I want to turn to the second force that shapes decision making and use of natural resources: raiding, violence, and the threat of raiding. Raiding among East African pastoral peoples has been the subject of much debate and discussion, especially concerning its motivations and impacts. The topic of intertribal and intratribal hostilities was one of the central themes in the first major anthropological study of an East African pastoral people, *The Nuer* by Evans-Pritchard (1940). Since then numerous articles and book chapters have been written about the role of raiding in East African pastoral societies.

The debate concerning warfare and raiding among East African peoples is one aspect of a much larger literature that is geographically global, covers a million years of human history, and questions the very nature of human beings. Before addressing the East African context in more detail, I feel it would be useful to provide a brief summary of some of the larger issues and debates within which the Turkana and other East African material can be contexualized.

WARFARE AND RAIDING OUTSIDE OF EAST AFRICA

Questions about why people have wars and kill each other have been the subject of social science inquiry and debate for over a hundred years. In the past many researchers have often viewed raiding and warfare among small-scale societies as an aberration, only to be mentioned in passing. Others have viewed warfare as constituting one of the most important influences in the development of complex societies. According to Carneiro, "Warfare has had a profound effect on human history—indeed, it has been the principal means by which human societies, starting as small, simple autonomous communities, have been transformed, step by step, into vast and complex states and empires" (1994:5).

The evolutionary perspective that frames Carneiro's discussion harkens back to the early part of the twentieth century when social scientists incorporated raiding and warfare into their ladder of cultural evolution. Of course, at that time, warfare was not viewed as a process, but the type of war was one characteristic of the evolutionary stage of a particular culture. "Customs, practices, or weapons were placed in sequences or they were linked or related to stages of an evolutionary typology, such as levels of subsistence technology" (Otterbein 1999:795). Over the course of the next eighty years the theoretical frameworks used to frame discussions and explain behavior related to war and human aggression increased dramatically; for example, Otterbein identifies sixteen theories or approaches being debated during what he calls the "Classic Age" (1960–80).

Conflict and Controversy

Today the debate continues, fueled by current accounts of ethnic genocide and accusations of ethical misconduct on the part of researchers that may have exacerbated the level of conflict that they were supposedly studying. The practice of warfare, raiding, and violence has always precipitated passionate discussions. In the words of Anna Simons: "War is a fraught subject. Those who study it often fight about it" (1999:74). She goes on to cite the writings of Hallpike (1973), Ferguson (1996), Chagnon (1997), and Keegan (1997) as examples of the kind of invective found in this body of literature.

Chagnon's work among the Yanomami, in particular, has served as a lightning rod for much of the criticism. According to Sponsel: "There has been a disproportionate amount of controversy, debate, and criticism, some surprisingly aggressive and personal, surrounding Chagnon's characterization of the Yanomami as 'the fierce people' as well as around his sociobiological explanation of their aggression and other matters" (1998:98). What was primarily an academic discussion became distinctly personal with the publication in 2000 of *Darkness in El Dorado* by Patrick Tierney.

In this book Tierney accused geneticist James V. Neel of possibly causing, and certainly exacerbating, a measles epidemic among the Yanomami by deliberately administering an out-of-date vaccine. He also charged Chagnon with misrepresenting the Yanomami people and promoting intertribal violence. In the words of Fernando Coronil, "Tierney argues that Chagnon created the myth of the Yanomami as the 'fierce people' through his own brand of physical and symbolic violence against them . . . and used his power and material resources to obtain information, often through bribes and coercion, about personal names and genealogies (which are taboo to reveal), created divisions by distributing valuable goods among different factions, promoted warfare for film performances, and misrepresented the Yanomami as extraordinarily violent people" (2001:265–66).

In the fall of 2001, following months of debate, the American Anthropological Association set up a task force to investigate Tierney's claims. Despite Sponsel's claim that "Tierney exposed the ugliest affair in the entire history of anthropology" (2002:149), the task force found that most of Tierney's accusations were unfounded. We have not heard the last of this, but this illustrates the rancor surrounding aspects of debate concerning the nature of war, raiding, and violence.

The Peaceful Savage?

This debate has also highlighted one of the more important divisions among researchers: those who believe that warfare, raiding, and violent aggression are part and parcel of the human condition—part of our biological makeup and our evolutionary past—and those who believe that humans are basically peaceful, but that warfare or raiding may result from a particular set of circumstances or conditions at a particular time. Some have characterized this division as those who believe in the "myth of the peaceful savage" versus the vision of Thomas Hobbes that characterized life as "nasty, brutish, and short" within a context of "war of all men against all men" (Hobbes, quoted in Whitehead 2001:835).

Otterbein has written about this division, classifying researchers into the categories of hawks and doves (1999). He considers the hawks as including anthropologists Lawrence Keeley and Robert Carneiro; military historians Arther Ferrill, John Keegan, James McRandle, and Robert O'Connell; primatologist Richard Wrangham; and biologist Barbara Ehrenreich. Doves include anthropologists Brian Ferguson, Neil Whitehead, Leslie Sponsel, Elman Service, C. R. Halpike, and Thomas Gregor; political scientist Richard Gabriel; and archaeologist Jonathan Hass. This categorization has been criticized as both over-simplifying and misrepresenting the writings of both hawks and doves. "Some of the individuals that Otterbein labels as 'hawks' are apologists for war, or at least their research can be used or misused that way. Otterbein's labels 'doves' and 'hawks' are a legacy of the politics of the Vietnam War. Some twenty-five years later, such loaded words can only serve to prejudge and greatly oversimplify the positions of those Otterbein names as such" (Sponsel 2000:838).[1] I agree with Sponsel's critique, but this classification does allow some-one unfamiliar with the literature some insight into one of the major issues of debate.

Those researchers whom Otterbein classifies as doves do not nec-essarily believe that humans live in harmony with one another. Rather, they believe that warfare and raiding are rather recent intro-ductions in our cultural evolution, and in the case of intertribal war-fare, the direct and indirect impacts of Western influence have some-times caused, and frequently exacerbated, the level and degree of conflict. This perspective is countered by sociobiologists like Chagnon and by E. O. Wilson (who has also written extensively about the Yanomami), and supported by archaeologists Keeley and Walker and primatologists Wrangham and Peterson. The sociobiologists argue that humans, especially men, seek to gain a reproductive advantage over other individuals and groups, and that raiding and warfare are a means to achieve this goal. Lawrence Keeley, in his book *War before Civilization* (1996), argues that indigenous peoples engaged in warfare long before any contact with Western societies. Although Keeley has been criticized for his "particular view of the discipline" (Whitehead 2001:834), his argument is supported by Phillip Walker's bioarchaeological research. Based on the examination of skeletal remains from around the world, Walker has concluded that

> Throughout the history of our species, interpersonal violence, especially among men, has been prevalent. Cannibalism seems to have been widespread, and mass killings, homicides, and assault injuries are well documented in both the Old and New Worlds. No form of social organization, mode of production, or environmental setting appears to have remained free from interpersonal violence for long. (2001:573)

THE TRIBAL ZONE

The origins of warfare and the degree to which it can be explained by Darwinian theory are among many issues and questions raised by researchers in the examination of raiding, warfare, and violence. The literature is vast, spans many disciplines, and as has been illustrated here, is often conflicting and contradictory. A number of edited books have been published during the last fifteen years that have either directly examined warfare among "tribal groups" or included the impacts of outside forces on local and tribal conflicts (see, for example, Rubinstein and Foster 1988; Hass 1990; Ferguson and Whitehead 1992; Reyna and Downs 1994; Turton 1997).

However, many of the debates have revolved around two ethnographic cases: the Yanomami and tribal warfare in New Guinea. The case of East African pastoralists has also been well documented but has not generated as much the theoretical discussion as the first two ethnographic areas. For the Yanomami, Sponsel (1998) has written an extensive review, as has Ferguson (2001). Sometimes the same data sets are interpreted completely differently, as in Chagnon's interpretation that successful warriors have more children (1988) and Ferguson's argument that the same data demonstrate exactly the opposite (2001).

Yanomami

The ethnographic literature for the Yanomami is so large that the study is sometimes referred to as Yanomamology (Sponsel 1998). More than sixty books and hundreds of articles have been written about the Yanomami, most of which examine various aspects of warfare or aggression. A number of researchers have spent long periods of time among the Yanomami, including Chagnon, Lizot, Ramos, Good, and Albert. Ferguson has written some of the better known theoretical examinations of Yanomami warfare but has never conducted fieldwork among them.

Sponsel (1998:100) has listed four different explanations for Yanomami aggression: "eclecticism (Chagnon 1968), cultural materialism (Harris 1984a, 1984b), sociobiology (Chagnon 1988, 1990), and regional political economy during colonialism (Ferguson and Whitehead 1992; Ferguson 1995)." He feels that Lizot's work (1985, 1994) only proved an "enticing hint" (Sponsel 1998:100) of a mentalistic explanation for Yanomami aggression and warfare.

Ferguson classifies the Yanomami literature, at least that written in the last thirty years, into three broad categories: materialistic explanations, cultural approaches, and biological perspectives (2001:99). Materialist explanations "see war as following self interest, an effort to maintain or improve material conditions" (99). Ferguson's analysis of Yanomami warfare is both materialistic and historical. His major argument is that "wars break out at moments of major change in the Western presence" (100). The Western presence in this case was primarily access to steel tools, and sometimes shotguns, that resulted in changes in both agricultural production and the balance of power among villages. Illustrating the kind of invective that has characterized the debate between Chagnon and Ferguson is Chagnon's 1996 review of Ferguson's book *Yanomami Warfare: A Political History*. Chagnon summarized Ferguson's argument as follows.

One might puckishly characterize his overall argument as the Amazon Basin equivalent of *The Gods Must Be Crazy* theory: the Kalahari natives were peaceful until a Coke bottle was discarded in their midst by a careless bush pilot and then war broke out. Instead of a Coke bottle, it is the machete or ax in the Amazon. Instead of a bush pilot, it started with slavers and rubber tappers, but more recently it has been missionaries and anthropologists, the most notorious has been the reviewer. (Chagnon 1996:670)

Ferguson considers "cultural explanations" as ones that include interpretive and hermeneutic approaches and says that they "stress that goals, conduct, and meaning should be seen within the logistics of broader cultural understandings" (2001:104). He views the work of Lizot (1994) and Albert (1989, 1990) as making the strongest contributions in this field within the Yanomami context. He also feels that the materialist explanation and the cultural explanation can be complementary to one another. On the other hand, Ferguson feels that the biological or sociobiological explanations for Yanomami warfare have no potential overlap with the other explanations. Chagnon, of course is the leading exponent of this theoretical position, but Ferguson also cites the work of Daly and Wilson (1988), Hrdy (1999), Wrangham (1999), and Wilson (1978) as bearing on the sociobiological explanation for Yanomami warfare. Although he has major difficulties with the theory itself, with respect to the Yanomami literature he rejects this approach because "the biological explanations are so consistently directly contradicted by the Yanomami data" (2001:112).

New Guinea

The literature on warfare among the indigenous peoples of highland New Guinea is also extensive, but the theoretical debates are not as caustic or personal as those described previously. Tribal warfare in New Guinea was popularized with the documentary film *Dead Birds*, made in 1963. In this portrayal, the Dani were seen engaging in "ritual" warfare, described by Shankman as involving "hundreds of men on each side of a designated battleground firing arrows in a highly individualistic fashion. The rationale for ritual warfare is revenge in order to placate the ghosts of the dead. These frequent wars are generally inconclusive and causalities are low" (1991:301). This is opposed to "secular" warfare, which is deadly, results in territorial realignment, but is rare.

Much of the debate concerning New Guinea warfare has contrasted ecological versus nonecological explanations as the underlying cause for war. Knauft (1990) has published a review of this literature, and articles by Shankman (1991), Pospisil (1994), and Strathern (1992) and books by Meggitt (1977) and Feil (1987) provide a good introduction to the literature. Shankman lists supporters of the ecological explanation as the cause of New Guinea warfare as Rappaport (1984), Vayda (1976), Meggitt (1977), and Morren (1987) and those opposing an ecological explanation as Koch (1974), Hallpike (1973), Sillitoe (1977, 1978), Feil (1987), and Vayda (1989).[2]

Rappaport's *Pigs for the Ancestors* (1968, 1984) argues that warfare and ritual are the means by which the human population rebalances demographics and natural resources in order to remain in equilibrium with the ecosystem. As mentioned in chapter 2, it is one of the most important books ever written in human ecology. But the literature is more complex than the ecological/nonecological debate. With respect to the New Guinea material Simons states:

> Even a summary reading of a fraction of the recent literature could leave the impression that the Dani may or may not fight to gain access to material goods (Blick 1988, Shankman 1991), whereas the Avatip fight in order to create community (Harrison 1989) and the Chimbu fight from having lost community (Podolefsky 1984). . . . Melanesianists writing about warfare use three different frames of analysis and begin their examinations at different (but still connected) sociotemporal points. The local ecology provides certain constraints and opportunities (Vayda 1971; Rappaport 1968; Ember 1982); the local culture into which people are socialized creates constraints and opportunities (Sillitoe 1978; Tuzin 1997; Roscoe 1996); and supralocal events and institutions have constrained as well as liberated indigenous societies (Knauft 1990). (1999:81–82)

Beyond the Tribal Zone

Although warfare in New Guinea and warfare among the Yanomami have dominated the theoretical discussions concerning warfare and raiding in small-scale societies, many researchers today are examining modern warfare and ethnic genocide. Sponsel mentions the work of Nordstrom (1995, 1997) on Mozambique, Montejo (1987) on Guatemala, Holtzman (2000) on Nuer refugees in Minnesota, and the anthologies edited by Fry and Bjorkvist (1997), Nordstrom and Robben (1995), and Sluka (2000) as among the best of this new genre (Sponsel 2000). Simons provides a far more extensive list of social scientists who have recently examined regional and ethnic conflicts (1999:85). While the theory of war and its causes remains important, some of the more recent literature addresses practical issues such as terrorism, human rights, and conflict resolution.

In returning to the ethnographic context of East Africa and the Turkana, many of the discussions and debates described earlier will prove relevant. Intertribal raiding and ethnic conflict, and the extent to which they are influenced by the state, have been a major topic of debate (discussed later). Relationships among ethnic groups are complex, negotiated and renegotiated as environmental conditions and political contexts change.

WARFARE AND RAIDING AMONG THE TURKANA

Among the Turkana, it is difficult to underestimate the degree to which raiding and violence influence people's lives. At any time a herd owner could lose all his livestock, and members of his family could be injured or killed. Even during times of relative peace, the topic of potential raids by enemies is a frequent topic of men's discussion while sitting under the tree. Throughout the year scouts are sent to unused water holes to look for footprints or other evidence that enemies are nearby. Women know that they are especially vulnerable to enemy raids, as often many of the men are separated from the major awi. Attacks can be vicious, brutal, and deadly, often coming just before dawn with little or no warning.

For the Ngisonyoka, the major threat comes from the Pokot, living to the south and west. For other Turkana sections the enemies lie to the west (the Karamojong, the Jie, the Dodoth) and to the north (the Taposa and the Dassanetch). The Samburu live to the east, across the Suguta valley, but they are not viewed as a serious threat; instead they are often the subjects of Turkana raids. The Turkana have a reputation as being fierce warriors and deadly enemies. In the Turkana language the word for stranger and the word for enemy is the same, *emoit*, and as a general rule all enemies should be killed. However, during the period covered in this study, 1980 through 1996, the balance of power between the Turkana and ethnic groups bordering Turkana District shifted against the Turkana. To the north, the Sudanese government provided sophisticated arms to the Taposa, in the hope that they would use them against the Southern Peoples Liberation Army (SPLA). Some Taposa may have fought against the SPLA, but the guns were also used to stage raids against the Turkana. To the south the Pokot were able get access to numerous automatic weapons, particularly Kalashnikovs (AK-47s), aided by local politicians and other members of the Kalenjin faction that controlled the national government.[3] In a 1994 article published in the British newsmagazine *The Economist* (July 16, p. 38), it was estimated that over 10,000 Turkana had been killed in raiding in the 1991–94 period alone.

Despite the occasional story, it is generally difficult to get information about raiding once outside of Turkana District. National newspapers only carry stories on raiding when the raids are large enough to be shocking or might result in an international incident. This is even more true for international news agencies. Nevertheless the following headlines from East African media demonstrate the violence associated with raiding in Turkana District.

Turkana District-Massacre at Dawn: 400 cattle rustlers kill 192 in an orgy of violence (*Weekly Review,* April 22, 1988)

71 died in Lodwar attack (*Daily Nation,* July 13, 1993)

37 raiders killed in border clash (*Daily Nation,* July 17, 1993)

The death toll from Saturday's raids in the Lorengipi and Lokirima areas of Turkana District has risen to 46 (*Daily Nation,* Sept. 23, 1997)

Slaughter in Turkana: Forty people are killed following a raid on the Turkana by Pokot cattle rustlers (*Weekly Review,* Sept. 26, 1997)

Death toll rises to 100 in stock clashes (*Daily Nation,* March 9, 1999)

Killings anger Turkana leaders (following a raid by Pokot in which 26 Turkana were killed) (CNN, April 2000)

The object of the raids is to capture livestock and to kill enemies. For each of the raids mentioned in the newspapers, tens, perhaps hundreds, more occur. They are generally smaller and may involve only the taking of livestock, or only a few people may be killed. The violence associated with the raiding seems to have escalated in recent years, and in this regard it is worth quoting a few lines from one article.

> The death toll in fighting between the Pokot and Turkana herdsmen has risen to 100, a church leader said yesterday. Among those killed were 28 women believed to have been raped in the manyattas before being shot, said Bishop Stephen Kewasis.
> The Anglican Bishop of Kitale said he had personally counted 100 bullet-riddled bodies at the scene of the fighting—Naromoru centre in Turkana. At least 15 of those killed were children aged below 12, said Dr. Geoffrey Kasembeli of Lodwar District Hospital. (*Daily Nation*, March 9, 1999)

Raiding not only results in the loss of human lives and livelihoods but also exerts a very strong influence on how the land is and is not used. Border areas, regardless of resources available there, are often viewed as no-man's-land with little or no human presence. Areas close to border areas are avoided during times of hostilities unless environmental conditions are such that there is little or no choice but to move into these areas. Although the underlying motivation for raiding has been hotly debated, the case study that follows demonstrates that such motivation cannot be reduced to a single factor but lies in the articulation of the political, social, and ecological factors, as well as events at particular times and in particular places.

Guns and Globalism

Guns have historically provided the connection between those living in and around Turkana District and the outside world. For the last decade, anthropologists and other social scientists have been examining the impact of global forces on local peoples. Even in some of the more remote places, and among people who appear, at least on the surface, to be stereotypically traditional, local livelihoods and social practices have been influenced by colonialism, global markets, and political agendas for many decades. For example, Hutchinson's book on the Nuer examines how "global historical forces," especially state involvement and the war in the southern Sudan, have transformed social power relations and challenged some of the values that Evans-Pritchard considered to be at the core of Nuer society (Hutchinson 1996). Piot's recent book *Remotely Global* (1999) argues eloquently that colonialism and postcolonial influences have shaped culture and everyday life for the Kabre of Togo.

Clearly the availability of guns and more recently Kenyan, Ugandan, Sudanese, and Ethiopian politics have impacted local people in Turkana and the way they use their land. Whereas Hutchinson's and Piot's studies emphasize the influence of "global" forces on the social and symbolic lives of local people, here I examine how these forces influence the way people use the land and how they manage their natural resources. This relationship of outside forces articulating with local livelihoods and the use of natural resources fits well with the emerging paradigm of political ecology. Disequilibrial ecosystems, political instability, and differential access to modern weapons can be a volatile combination, and all need to be considered in order to understand raiding and violence in Turkana.

The Regional Context of Raiding

Stories about the "aggressiveness" of East African pastoral people go back to the accounts of early European explorers. The so-called blood lust of the pastoral inland tribes and their need for "trophies" posed a major obstacle for trade from the interior to the East African coast in the late nineteenth and early twentieth centuries. Among those pastoralists singled out as especially aggressive and warlike were the Maasai. However, some scholars such as Alan Jacobs feel that the stories of Maasai aggression were just that—stories, designed to limit the safari trade inland and allow it to be controlled by a handful of Swahili traders (Jacobs 1979). It was certainly true that following the devastations resulting from rinderpest and smallpox many Maasai engaged in inter-sectional warfare, but on the whole the image of the overly aggressive Maasai pastoralist may have been based more on fiction than fact.

In the literature concerning warfare and raiding among East African pastoralists, two publications stand out. Both are edited volumes, and both were the results of a conference held in Japan: *Warfare among East African Herders,* published in the Senri Ethnological Series (Fukui and Turton 1979), and *Ethnicity and Conflict in the Horn of Africa* (Fukui and Markakis 1994a). More recent publications have addressed the role of the state and civil unrest and its impact on local peoples. Examples of this research include Hutchinson 1996 for the southern Sudan and Hendrickson, Armon, and Mearns 1996, 1998 for the changing nature of raiding in Turkana District.

Raiding, counterraiding, and warfare have been characteristic of the people living in northern Kenya, northwestern Uganda, southern Sudan, and southern Ethiopia for much of the twentieth century. This area has sometimes been referred to as "the tribal zone" because it lies at the margin of state influence (Fukui and Markakis 1994b:2). However, as Hutchinson (1996) and Hendrickson, Armon, and Mearns (1998) among others point out, the state has exacerbated both the degree and intensity of conflict in these areas. According to Fukui and Turton (1979), conflict among groups in this area began to increase beginning around 1970, precipitated by severe drought, famine, and the inflow of firearms. From then on the situation only deteriorated. In the words of Fukui and Markakis:

> Much worse was in store for the region in the 1980s, when drought became the rule and famine reached biblical proportions. War spread to agricultural regions, and was fought on many levels simultaneously—between states, regions, ethnic groups, clans, and lineages. Thanks to sophisticated weaponry generously supplied by patrons from abroad, warfare was waged on high technical levels far above the region's native capacity, and the antiquated Mannlicher rifle, the weapon of the 1970s, was now replaced by the Kalashnikov (AK-47). Such escalation changed the nature of war just as anthropologists were coming to grips with the subject, and some of them found the groups they were studying facing extinction. (1994b:2)

Raids against some of the smaller groups challenged their survival as an independent ethnic group. Tornay has described how the Nyangatom lost approximately 10 percent of their population in the early 1970s, mainly due to raids by the Dassanetch (1979). Turton has described how the Nyangatom reemerged as a power in southern Ethiopia through access to automatic weapons, and how they in turn launched a series of devastating raids against the Mursi, forcing them to contemplate the possibility that the survivors would be scattered and no longer be able to identify themselves as Mursi (1994).

Recent articles on the Karimojong have demonstrated how the large influx of guns over the last twenty years has impacted that society (Gray et al. 2003; Gray 2000; Mirzeler and Young 2000). Mirzeler and Young (2000) cite Ugandan government sources that estimated the number of AK-47s (automatic rifles) held by the Karimojong and their neighbors at 30,000 to 40,000, while Gray (2000) cites rumors that suggest that the number may be as high as 100,000. Mirzeler and Young argue that the increase of guns has disrupted the relationship of the Karimojong to the state, transformed some aspects of social organization (a decline in the influence of elders and the emergence of warlords), and developed informal markets in remote border areas where guns and ammunition are sold.

Gray and her colleagues see the proliferation of automatic weapons and the increase of raiding as having an even larger impact on the Karimojong.

> By disrupting seasonal migration routes, by constraining subsistence activity, by forcing the closure of trade routes in and out of the district, by altering the structure of marriage and the stability of social networks, and by interfering with health care delivery, armed cattle raiding emerges as the critical factor in recurring crop failures, herd losses, food shortages, disease outbreaks, and sustained high mortality since the 1970s. (Gray et al. 2003)

Motivations

The motivation for raiding has been one of the enduring themes in this body of literature. To some extent the discussion has followed major topics of debate in the larger anthropological literature. Materialist versus nonmaterialist, ecological versus economic, and structuralist versus poststructuralist explanations have all been put forward and debated. Turton (1994) begins his argument for raiding among the Mursi with a discussion of the Hobbesian understanding of human nature, the role of political institutions, and how this had been incorporated into Durkheim's view of society and the individual. He argues that raiding, at least for the Mursi and their neighbors, is the means by which their identity as an independent political unit is "created and kept alive" (23). Authors such as Fukui and Markakis (1994) and Bollig (1990) have argued that individual agency provides the underlying explanation for raiding that could be economic or social or incorporate elements of both.

Ecological Explanations

To a large extent, ecological explanations have revolved around two main issues: territorial defense and recuperation from losses stemming from drought and disease. Most authors agree that, in general, the primary goal of raiding is not the occupation of enemy territory; examples of this occur but are rare in the ethnographic literature. However, gaining seasonal access to grazing lands and water holes and defending these resources from incursion have been used as a conceptual framework to explain raiding (R. Dyson-Hudson and Smith 1978; McCabe 1990).

Large-scale fluctuations in the livestock populations due to recurrent droughts and occasional outbreaks of epizootics are characteristic of arid lands and nonequilibrial ecosystems. Although not phrased in terms of ecosystem dynamics, the need to restock following stressful periods has often been used to explain the motivation and function of raiding. Dietz mentioned that raiding among the Pokot dramatically increased following two consecutive years of drought (1987), while Markakis found that "scarcity and mobility make conflict inevitable" (1994:219). Cousins carries this argument further, making the link with an understanding of arid rangelands as nonequilibrial ecosystems (1996). Drawing on the work of Behnke, Scoones, and Niamir-Fuller he states:

> The "new thinking" [nonequilibrium models] also asserts that a situation of chronic or endemic conflict is a central feature of nonequilibrium settings (Behnke and Scoones 1993; Niamir-Fuller 1994; Scoones 1994). This helps explain the high degree of inter-group conflict often associated with pastoralism, but also the patterns of cooperation and reciprocal access which are found. Environmental variability thus results in a high degree of political (and sometimes military) competition, ameliorated by

periods when competitors relate to each other as allies, neighbors, or even kin (Behnke 1994:5). (Cousins 1996:43)

Social and Cultural Explanations

Social and cultural reasons have also been used to explain raiding
among East African pastoralists. Lamphear mentions the importance
of raiding to young men in Turkana (1992). He describes going on
raids as a rite of passage linked to prestige in the larger community.
Bollig (1990) has written about the high bride-price among the Pokot
and how this is related to raiding (e.g., by raiding young men can
acquire the livestock necessary for marriage that they may not be able
to acquire through inheritance). Many authors, including Baxter
(1979), Fukui (1979), and Turton (1979), mention the prestige associated with the killing of an enemy and how this encourages young men
to raid neighboring groups. The ways that people conceptualize
themselves and identify as a group also have been related to raiding,
particularly by Turton (1979, 1994).

Economic Explanations

Finally, economic explanations for raiding are more common today
than they were in the past. Bollig provides a short economic analysis
on the value of guns and raiding for Pokot individuals (1990). He
points out how an investment in an automatic rifle, even if the initial
cost is quite high, will pay rather large dividends in the long run.
Fleisher, in his analysis of raiding among the Kuria of Tanzania, also
advocates an economic explanation. He sums up his position as follows: "Kuria cattle raiders, however, are in it for the money, and have
been for approximately the last eighty years" (1999:238).

With respect to raiding against Turkana, Hendrickson, Armon, and
Mearns have discussed the transition from "redistributive raiding" to
"predatory raiding" (1998). Redistributive raiding occurs on a relatively small scale and serves to rebuild herds and redistribute animals
over a large area. Livestock are not sold, and conflict is constrained by
social norms and the influence of elders. Predatory raiding is a different thing all together. These authors view predatory raiding as
defined by

the growing involvement in raiding of actors outside of the pastoral system which has significantly undermined pastoral livelihoods and the socioeconomic integrity of the pastoral systems as a whole. . . . It is driven by a criminal logic contrasting sharply with former notions of balance and reciprocity. Predatory raids are largely initiated by people outside Turkana, including armed military or bandit groups in Kenya or surrounding states as well as "economic entrepreneurs." The motives are commercial: to procure cattle in vast quantities either to feed warring armies or to sell on the market for profit. (Hendrickson, Armon, and Mearns 1998:191)

Personal Observations

Although I am providing an overview of the literature here, I want to expand upon a few points based on my own research. I can attest to the prestige given to raiders among the Turkana. Like many of their neighbors, Turkana practice scarification upon killing an enemy. For the Turkana a band of scars across the right shoulder indicates the killing of a male enemy, a band of scars across the left shoulder signifies the killing of a female enemy. Livestock that are accumulated in raiding are often distributed throughout the extended family, and thus a son, brother, or son-in-law who is a successful raider is held in especially high regard. Young women will sing songs about young men who are successful raiders, and such individuals often receive the benefits of their adulation. A short vignette may also be appropriate here.

One day in 1987 I was sitting with a group of young men and women on a bench overlooking the street and shops in the border town of Kainook. Although the town was just inside Turkana District, the large army and police presence made it safe for Pokot to come into town and visit the shops. While we were sitting on the bench, a Pokot man walked by on his way to a shop up the street. One of the young women with us turned to the young men sitting on the bench and said: "There goes a Pokot—who's going to go kill him?" Following a short silence other women began to chime in: "Oh you're all big men when we are back at the awi, but now that you see a Pokot you're afraid to do anything." The young men hemmed and hawed and generally told the young women to stop giving them such a hard time. They knew that it was impossible to do anything while all the police and army personnel were around. Of course all of this was said in jest, but to me the underlying message was clear. Impressing and pleasing women, not just acquiring livestock, provide a strong incentive to participate in raids.

I also believe that there is a very personal aspect to raiding and the way individuals feel about other peoples. The Pokot had killed children, wives, brothers, and sisters of many of the Ngisonyoka whom I came to know well. One may be able to identify some underlying ecological or economic factors that help explain the perpetuation of raiding, but when members of another group kill those you love, the relationship becomes personal. Exacting revenge is a powerful motivation, and it is not easy to persuade those who have lost a loved one to pursue a course of peace. I do not think that I am exaggerating when I say many of my friends felt real hatred for the Pokot people. Nevertheless, they were willing to try to engage in peaceful relations, and I accompanied men to peace meetings on a number of occasions. I realize that this is a Turkana-oriented view; my impression is that researchers like Michael Bollig who worked with the Pokot felt similarly, but in that case the enemies and the killers were the Turkana.

Historical and Political Context of Raiding in Turkana

Much of the material relevant to the historical understanding of raiding on and by Turkana has already been covered (chap. 3). However, there are a few points that I want to emphasize with specific reference to raiding. The Turkana expanded at the expense of those already living in what is now Turkana District. The expansion involved incorporating communities, exterminating communities, and driving communities away from Eturkan (Turkanaland). By about 1850 the expansion had reached its spatial limit, and raiding was no longer necessary to acquire new lands. By the 1890s the Turkana had consolidated these gains. Two factors, guns and disease, both brought from outside East Africa, contributed to the Turkana expansion and consolidation of their territory. Guns came from the north, obtained from Ethiopian traders. Contagious bovine pleuropneumonia and, later, rinderpest considerably weakened many of Turkana's neighbors but had only minor impact on Turkana livestock.

During the early 1890s, the Turkana became embroiled in international politics and border disputes between Britain and Ethiopia. Up until the second decade of the twentieth century, the British did not have much direct involvement with the Turkana. They were engaged in protecting their interests in the more fertile areas to the south and the west. Their objective was to contain the Turkana and to keep them from causing too much disruption with the ethnic groups that shared the better watered highlands. The Turkana viewed the British as protecting and supporting their enemies and thus saw the British as their enemies also. The Ethiopians wanted to extend their southern boundaries into what is now Turkana District and allied themselves with the Turkana on raids against tribes such as the Dodoth. By 1914 the British decided that the Ethiopians needed to be completely pushed out of Turkana and that the strength of the Turkana needed to be broken. However, World War I postponed this confrontation.

The Labur expedition of 1918 accomplished both of these tasks. Of course the "British" army that conducted the expedition was mainly composed of African soldiers and fighters, many of whom came from ethnic groups that were at war with the Turkana. Lamphear mentions that prior to this in some border areas, particularly to the south, there was intermixing of ethnic groups to the extent that it was becoming difficult to distinguish Turkana from Pokot. Intermarriage, trading, and stock alliances among individuals were the order of the day. All this broke down as the degree and intensity of raiding increased. Thus during this period ethnic conflict was exacerbated by colonialism and international politics, which can still be seen today in many parts of the "tribal zone" described previously.

Following the devastation left by the Labur Patrol, a period of peace was ushered in that lasted until the 1960s, at least in southern Turkana. The Ngisonyoka again traded, formed individual stock alliances, and intermarried with the Pokot.

In the north, the Turkana were once again embroiled in international politics. In the period leading up to World War II, Italians in Ethiopia began supplying the Dassanetch with guns and encouraging them to attack the Turkana. The British maintained tight control of arms on the Kenyan side, and by 1939, over 250 Turkana had been killed by Dassanetch raiders. A combined British and Turkana force began patrolling the border in 1940, and the Dassanetch pulled back. When the war ended the British again disarmed the Turkana, but the Ethiopian government did not do the same with the Dassanetch. By the 1950s, the Dassanetch were again engaged in regular attacks on northern Turkana, killing hundreds of people, taking thousands of livestock, and forcing the Turkana to move south, away from the border.

Although the mid-1920s through the mid-1960s were characterized by peace between Turkana and Pokot, the Pokot were moving west and raiding the Karimojong and the Sebei in the mid-1950s. As mentioned earlier, Dietz (1987) attributes periods of Pokot aggression as being associated with two or more years of drought. The Pokot experienced drought from 1950 through 1953 and again in 1964 and 1965. In the early 1950s, the Pokot were beginning to raid into Karamoja frequently, and by the mid-1960s the Pokot and Karimojong were engaged in full-scale warfare. Dietz mentions that the Karamoja annual report for 1964 cites over 1,000 raids between the Karimojong and the Pokot (1987:125). In the mid-1960s the Pokot also began to raid against the Turkana, thus breaking the peace that had lasted for nearly forty years.

Hostilities between the southern Turkana and the Pokot rapidly escalated then, especially after a raid in 1967 on the town of Kaputir in which 33 women and children were mutilated and killed (Dietz 1987). Both sides were acquiring weapons at this time. It is also at this time that the first mention is made of Turkana raiders as *ngorokos*. Dietz describes them as "armed with guns instead of spears; stolen cattle was transported to the north, using lorries sometimes; women and children were killed too and also food stores and other property was destroyed" (1987:126). Bollig describes raids by the Turkana in the mid-1970s as being conducted by "Turkana tribal warriors and a very well organized group of Turkana bandits (Ngoroko)" (1990:74).

The years 1969 and 1979 are both named *Ekaru GSU* (year of the GSU) by the people of south Turkana. This refers to the disarming expeditions carried out by the elite General Service Unit of the Kenyan army. I have described these events earlier, but I want to reiterate here that this upset the balance of power in south Turkana. The Ngisonyoka lost most of their guns during these operations, while the Pokot only lost a small fraction of theirs. Bollig mentions heavily armed Turkana raiders attacking Pokot in the 1970s, but these were raiders from the north, and the GSU "expedition" had not disarmed the northern sections. This resulted in periods of intense raiding by the Pokot against the Ngisonyoka and other Turkana peoples living along their borders in both Kenya and Uganda.

When the Tanzanian army defeated the forces of Idi Amin in 1979, the Karimojong, Pokot, and Turkana were all able to arm and rearm themselves easily. The barracks at Moroto were broken into, and I heard stories of men and boys bringing donkey load after donkey load of automatic weapons and ammunition into Turkana. Dietz (1987) mentions the massive escalation of Karimojong raiding following their weapons windfall.

Ngoroko

The Pokot were described at the beginning of this chapter as the major threat to the Ngisonyoka. While this is true, Ngoroko were also a problem, at times stealing livestock, harassing and beating people, and occasionally killing someone. The media refer to Ngoroko as bandits, but they were responsible for a large part of the raiding carried out by the Turkana during the study period.

The Ngoroko were permanently organized as warriors living in the rugged and remote mountains in northern and western Turkana. Although their primary objective was to raid enemy groups, during the late 1970s and early 1980s they also harassed the Ngisonyoka by demanding livestock to be eaten while they were encamped in their territory. It was customary for local people to help feed raiders on their way to attack the emoit, and it was customary for successful raiders to give back livestock, with interest, to those who fed them on their return journey. Because the Ngisonyoka had suffered serious reprisals at the hands of the government during the 1970s, people were reluctant to help in any way that could be interpreted by government officials as contributing to raiding. The northern sections of the Turkana were not disarmed during the GSU expeditions of 1969 and 1979, and when the Ngoroko arrived in Ngisonyoka country, they came very well armed. They also viewed members of the southern sections as cooperating with the government, something a "true" Turkana would not do.

During the early 1980s it was not unusual for a small group of Ngoroko to show up at water holes and demand a steer or a few goats from herd boys. However, by the mid-1980s Ngoroko began to take animals from homesteads and demanded to have sex with the teenage girls. Relationships deteriorated further after a group of Ngoroko stole some milking camels from an awi; on another occasion Ngoroko attacked a series of homesteads, breaking one man's arm and throwing a baby into a fire (luckily the baby survived). The only death that I know of resulted from an argument over a goat between a Ngoroko and the brother of one of our research assistants. He was shot trying to keep the Ngoroko from leading the goat off to be slaughtered.

I have heard two explanations about why these raiders or bandits were called Ngoroko, both having to do with the Turkana term for alternating black and white stripes. One explanation was that you could never tell a Ngoroko—he could be your friend one minute and enemy the next. The other was that a Ngoroko could be one way during the day, but turn into an enemy at night. These explanations referred to local people who would sometimes join the Ngoroko while they were nearby trying to keep their identities hidden, then would resume a pastoral life-style when the Ngoroko moved away.

The origin of the Ngoroko is not clear, but based on Lamphear's historical analysis of the Turkana it seems that the Ngoroko are a further development of the group that he refers to as "Ngomoroko." He describes them as "a more or less permanent force of men well armed with rifles, who stayed in the western mountains beyond the reach of the government." He goes on to say that they "became the ultimate heirs of the Ruru military tradition by the 1950s" (1992:253). The Ruru arose as a military organization in response to the British attacks during the second decade of the twentieth century. They were renowned for their courage, military organization, and marksmanship. However, Lamphear notes that by 1921 they had begun to act as a force unto themselves. His description of their "misdeeds" sounds remarkably similar to what I witnessed during the 1980s. "They would take cattle from people and slaughter them; they would rape people's wives and daughters. This caused them to be cursed by the elders.

People began to call them *Ngiakitiba* (Those of the Elders' Curse)"
(1992:227).

Drought, Raiding, and Famine

Although international attention has only sporadically been focused
on raiding in the Kenya, Ethiopia, Uganda, and Sudanese pastoral
zones, coverage of famine in the areas has been substantial. Drought
is usually considered the ultimate cause of the famines in this region,
but poor range management and civil war are often seen as contributing factors. Less attention has been paid to the role of raiding. According to Hendrickson, Armon, and Mearns (1998), three factors contributed to this lack of attention. First, there is a general ignorance
about pastoral peoples combined with an assumption that drought is
inevitable. Second, the tendency of Western aid agencies to view
famine as a unique event places emphasis on the "consequences of
famine rather than the cause" (1998:189). The final reason is political.
Here they quote Cullis and Pacey:

> Drought is politically neutral, and to present it as the cause of
> crisis avoids blaming national governments or district adminis-
> trators for failure to control the security situation. It also avoids
> the need to identify and remedy other factors which may
> encourage raiding—factors that might include chronic poverty,
> alienation from national institutions, and trading in weapons.
> (Cullis and Pacey 1992:8, quoted in Hendrickson, Armon, and
> Mearns 1998:189)

The Turkana situation is a classic example of how raiding and
famine are intertwined among East African pastoral peoples. A major
impact of raiding is that there are often areas that are only infrequently used because of the security risk, and these areas often have
among the best forage resources locally available. Ecosystems Ltd.
(1985) conducted an extensive study and analysis of land use in
Turkana district in 1982 through 1984. They found that approximately
47 percent of the land was unused, principally due to insecurity. My
own study found that about quarter of Ngisonyoka territory was
rarely used for the same reasons (McCabe 1990).

For those raided, the loss of livestock can be significant. Hardly mentioned in the literature is the fact that the type of livestock lost may be more important than the numbers of animals. If a family loses its milking animals, as happened to Angorot in 1996, hunger would be the immediate impact; the family's eventual loss of its pastoral livelihood may be the long-term result. Without milk, a herd owner has three options: he can sell animals to buy grain; he can slaughter animals; and he can borrow milking animals from friends and relatives. Obviously the third option is preferable, but in times of drought it is difficult to borrow milking livestock. Selling animals or slaughtering livestock is a short-term solution but can endanger the herd's ability to recover quickly. The conflict between short-term solutions to problems and long-term strategies is another tension confronting herd owners in times of stress, when the welfare of the family and that of the herd may both be at risk.

Another aspect of raiding and famine has to do with land use patterns. When large areas are considered off-limits during times of insecurity, people and livestock tend to cluster together for protection; often this involves many members of an extended family. During my 1985 study of famine in northern Turkana, I found that if this group was raided, then individuals were left with no one to turn to for help. The entire support group was made destitute in the course of a single raid.

Conclusions

Just as an understanding of the environment as a persistent but non-equilibrial ecosystem has provided the ecological context for the case study, an understanding of the importance of raiding and violence frames the historicopolitical context. The next chapters describe the families that I lived and worked with during the 1980s. I also describe the movement of the homesteads and of the herds, as well as analyzing what factors contributed to the decisions that were made. All decisions took into account the environmental condition of the rangelands and the availability of water. All decisions also took into account the security of the family and of the herds. Sometimes decisions were easy: moving to areas where forage was abundant, water was available, and the security risk was minimal. More often the opposite was the case. Movement into ecologically desirable areas involved putting both the family and livestock at risk. This tension, to a large extent, dominated the lives of the people and provides the theoretical context for the rest of the book.

Finally, I want to comment on my choice of title for this book and for this chapter: "Cattle bring us to our enemies." This is a quotation from a woman I interviewed during the late 1980s concerning individual preference for particular livestock species. I was interested in exploring the differences in how men and women viewed livestock. I asked individuals to rank-order cattle, camels, sheep, goats, and donkeys in terms of the animals that they liked the most. Surprisingly, many women ranked cattle quite low, and when I inquired about this people often answered that cattle did not give much milk, especially in the dry season. This of course made perfect sense, but one woman's answer moved the discussion to a whole different level. Her response at once captured the essence of the tension between environmental constraints on the one hand and the danger of raiding on the other. She had ranked cattle as her least-liked species, and when I asked her why, she said, "Because cattle bring us to our enemies." Cattle need grass, and as has just been pointed out, during most times of the year grass can only be found in the southern part of Ngisonyoka territory and in the highlands. In very bad years, grass will only be found in the southern highlands; this is the land that forms the border area with the Pokot. During the sixteen years that I worked with the Ngisonyoka I spent a good deal of time in this border area. The tension is palpable during the time when the families move here. Often women and children refuse to sleep in the awi, fearing night attacks by the Pokot. Environmental conditions and the viability of their pastoral livelihood draw them into the border zone; their safety and that of their livestock pushes them out. Cattle do indeed bring the Ngisonyoka to their enemies, as they do for the Nuer and many of the other pastoral peoples who migrate with their herds in this very dangerous part of the world.

PART 2

CHAPTER 6

Introduction to the
Four Families

This chapter introduces the four families with whom I worked for most of the time that I was in Turkana District. Much of the analysis concerning decision making is based on discussions and interviews that I had with the four herd owners who were the heads of their individual families. Men, therefore, have influenced my interpretation of events, in particular by how these men who make decisions concerning land use see the world around them. I also established friendships with a number of Turkana women and had many conversations with them about land use and family life. Where I can, I discuss what I perceived to be their impact on decision making. However, I do not pretend to present a balanced account of life among the Turkana. Mine is primarily a view from the perspective of Turkana males, for better or worse.

My goal here is twofold: to set the stage for the description of movement patterns and decision making in the analytical chapters that follow and to describe the four herd owners and their families so that the reader can get a sense of who they were as people, not just as analytical units or data sets. I previously provided historical detail on raiding so as to set the political and historical context of Ngisonyoka life when the research began in 1980. Here, I describe the families and livestock holdings of each of the four herd owners at the time that I first met them. I will also discuss some of the important personal events that occurred in the lives of the herd owners that led up to their social and economic conditions as of the time I arrived in 1980.

ANGOROT

Angorot is perhaps the most intelligent and articulate man I met during my entire time working with the Turkana. When I first met him he was in his early forties and living with three wives (one of them inherited), four brothers, and his mother. His herds consisted of 107 cattle, 68 camels, 722 goats, and 269 sheep. Although this number of livestock is considered a large holding, Angorot was not deemed to be a rich man. Angorot's fifth brother, Aki, was in prison when I first started working with him. The Kenya government disarmed the Ngisonyoka in 1979, and Aki was caught with a rifle. He was released in 1982 and returned to the awi, where he remained throughout the study period. Akopenyon was the principal manager of the camel herd and would accompany the nonmilking camels when they were separated from the awi. Nakwawi and later Lokora took care of the goats and sheep, and Erionga took care of the cattle. Before being sent to prison, Aki was in charge of the cattle, and after his return he took care of cattle when they were at the awi, but Erionga managed them when they were separated. Two features seem to characterize all of Angorot's brothers: a strong sense of responsibility and their playful sense of humor. Aki, for instance, loved to conceal my tape recorder and record parts of a conversation when people were visiting, then play the tape back catching them off guard. He would laugh so hard that sometimes he would fall off his *ekicholon* (small stool). His other brothers were not the jokesters that Aki was, but they all loved to play small jokes on each other and sometimes visitors (myself included).

In 1980 Angorot's awi consisted of six ekols. The first ekol belonged to Angorot's mother, Nakadeli. Angorot's unmarried brothers, Akopenyon, Nakwawi, Erionga, and Lorkora, also slept there when they were at the awi, as did Angorot's unmarried sister Kochodin. Nakadeli was the matriarch of the awi and offered advice and helped wherever needed. It was important that each of Angorot's wives get along with her as she clearly could influence Angorot in decisions related to the household. As is Turkana custom, each of the wives spent their initial years in the ekol of the household head's mother, living with and assisting her until the new wife had given birth.

The second ekol belonged to Angorot's first wife, Imadio, and her two children, a six-year-old son and a one-year-old daughter. Angorot's second wife, Kole, occupied the third ekol with her infant daughter. The fourth ekol belonged to Angorot's inherited wife, Lokaratch. Lokaratch was Angorot's father's fourth wife and brought four daughters, age nine to twenty, with her. Angorot also had an infant daughter with her. The fifth ekol belonged to Agule, who was Angorot's father's third wife. Agule chose not to be inherited and lived as an independent household head. One reason that she was able to do this was that she had an adult son who could manage the livestock. Also living in Agule's ekol were two other sons, age nineteen and eight, and three daughters, age twenty-one, thirteen, and eleven. Aki's wife Nateni lived in the sixth ekol with her infant daughter. In addition, four other young men, all cousins of Angorot, lived in the awi and assisted with herding tasks.

Angorot came from a quite humble background. His mother was her husband's second wife. When his father died in 1973, Angorot should have inherited all the animals that were allocated to his mother for milking, but instead Angorot's half brother Erionga,[1] the first son of the first wife, took almost all of the livestock for himself. Angorot began his career as a herd owner with only nine camels, three cows, and "a small fence of goats and sheep" (about thirty to forty). He married his first wife Imadio in 1971, and during the next two years she became pregnant twice but both pregnancies failed. Her first son, Lowoi, was born in 1974. Because Angorot only had a few animals following his father's death, he moved in with the family of his mother's brother and remained with him for approximately four years. During this time Angorot's herds grew. He acquired livestock through bridewealth payments when female relatives married, and he did not have to sell or slaughter many animals because his uncle was helping him with food.

In addition to bridewealth payments, Angorot was able to increase his herds by exchanging small stock for large stock, especially camels. Angorot grew up herding goats and sheep and based his herding strategy around the raising of these animals. As previously mentioned, small stock can give birth twice each year, and thus flock size can increase quite rapidly if carefully managed. The rapid increase in Angorot's livestock holdings demonstrate one aspect of the volatility of Turkana herd dynamics. Poor herd owners can become well off, possibly even rich, by the adept management of their herds. And, as will be demonstrated in the discussions of Lorimet and Lopericho, a relatively well-off herd owner can become impoverished almost overnight due to bad management or bad luck.

From 1980 to 1996 the size of Angorot's awi increased dramatically (chap. 9). All of his brothers married, some for the second time, and Angorot married two more wives. Angorot's recent wives were from the town of Lokori, and they continued to live there after their marriage. All the brothers remained together, pooling their resources but clearly under the direction of Angorot. Angorot's livestock holdings fluctuated quite a bit during this time. Livestock from Angorot's herds were used for bridewealth payments for each of the brothers. Angorot also lost many livestock to drought and disease in 1980–82, 1984–85, and 1990–93. In addition, the Pokot raided him four times during this period, costing him a total of 350 small stock, 58 cattle, 98 camels, and 15 donkeys. Details of these losses will be presented in later chapters.

To me, Angorot embodied all of the qualities that I came to associate with a good herd owner. A good herd owner is one that makes sound decisions concerning mobility and herd separation and that does not slaughter or sell animals without a very good reason (e.g., to feed children who are not getting enough milk). I have seen Angorot go for three days with no food, allocating his own milk to the children. Only when "the children were crying from hunger" did Angorot allow a goat to be killed to feed the members of the awi. In 1996 I asked Angorot's mother, Nakadeli, what Angorot was like as a boy. In typical Turkana fashion her answer was short, to the point, and related to livestock. She said: "He was a good herder."

The ecologists on the project were impressed with Angorot's curiosity and analytical capabilities concerning the environment. For example, during the early wet season of 1982, a large number of sheep were dying from an unidentified disease. When I asked Angorot what he thought was happening, he said he noticed when the sheep ate a lot of a particular species of forb they often became sick. We walked over to a patch where this plant was growing. He turned over some leaves and showed me a number of small white mites or insects on the bottom side of the leaves. He said: "See these little white bugs—I think that these are the things that are getting the sheep sick—not the

Fig. 6.1.   Angorot herding sheep

Fig. 6.2.   Angorot's mother Nakadeli in awi at dawn

Fig. 6.3.   Aki with son Terry

leaves." The exact cause of the illness was never determined, as Angorot soon moved the awi to a location where this particular species of plant did not grow. However, I rarely saw any other Turkana looking in such depth for the explanations for livestock disease or any other aspect of environmental relationships.

Angorot uses an aggressive strategy of herd management. He moves frequently and subdivides his herds more than most other herd owners. When large groups of people are moving south to gain access to better forage, Angorot's awi is often on the leading edge of the group. By doing so his animals will be able to eat the best forage, but this strategy exposes both the livestock and people to attack by enemies. Turkana often refer to the awis at the front of a group moving into dangerous areas as a "shield" for the other awis. As mentioned earlier, Angorot has lost substantial numbers of livestock to raids by the Pokot, although most of these losses were not the result of Angorot moving in front of other Turkana homesteads.

One reason that Angorot can employ such an aggressive management strategy is that he has abundant and reliable labor. He can depend upon his brothers to manage herds in a responsible manner when they are separated from the awi, and he has full confidence that the individuals involved in daily herding tasks are doing what is expected of them. The only time that Angorot experienced trouble with any of his brothers was in 1981. Erionga was put in charge of the cattle following the arrest of Aki in 1979. Angorot kept getting reports that the cattle, which were separated from the awi and located on the Loriu plateau, were being mismanaged and that Erionga was slaughtering livestock to supply meat for the people managing and herding cattle on the Loriu. Angorot had to make a trip to see his brother, and when he came back he told me that he was mad enough to beat Erionga. He said, however, he only reprimanded him and that he needed to trust him. Following this incident Erionga became an excellent herd manager and remains in charge of the family's cattle.

Lorimet

Lorimet is Angorot's brother-in-law but in many ways his opposite in terms of both temperament and personal history. When I first met Lorimet, his awi consisted of four ekols. Asekon, Lorimet's first wife and Angorot's sister, lived together with her infant daughter. Lorimet's mother, Natuk, occupied the second ekol, and Lorimet's brother, Achuka, stayed there when he was back at the awi. The second wife of Lorimet's father, Nangor lived in the third ekol. She maintained an independent household and separated from Lorimet about six months after I started working with him. Nangor lived together with her four daughters, ages twelve to twenty-one. The fourth ekol was occupied by Lorimet's married sister Napaipaia and her three daughters. Napaipaia's husband was arrested for possessing a gun during the same "operation" when Aki was arrested. Only one daughter was old enough to help with herding tasks; the others were five years old and an infant. She was therefore dependent on Lorimet for food, protection, and help with herding.

In 1980, Lorimet's livestock holdings consisted of 45 cattle, 30 camels, 231 goats, and 37 sheep. Unlike Angorot, Lorimet inherited a large number of animals following the death of his father, but his livestock holdings decreased as time passed. Natuk (Lorimet's mother) was the first wife of her husband and had three sons and one daughter. When Lorimet's father died in 1968, Lorimet inherited a herd consisting of approximately 60 to 100 camels, 25 cattle, and 100 to 150 small stock. Rather than setting out on his own immediately, he chose to live with his father's brother for five years before marrying and becoming a completely independent herd owner.

Lorimet has always had a problem with labor for his herds. In the mid-1970s Lorimet's brother, Lengess, left Turkana District and became a herder/laborer on a farm in the Kenya highlands. He had been shot in the leg during a Pokot raid and decided that living in Turkana District as a herder was just not worth the risk. However, he returned in 1987 and rejoined Lorimet's awi as the principal herder of the cattle. In 1994 he separated from Lorimet and became an independent herder.

Lorimet's other brother, Achuka, was a very unreliable herder and manager. When Lorimet separated his camels from the awi he had no choice but to put Achuka in charge of the animals. Achuka would often leave the camels in the care of other herders (very often Angorot's brother, Akopenyon) while he spent days in nearby towns drinking local beer and carousing. In 1983 he accompanied some Turkana bandits (Ngoroko) on raids against Ngisonyoka households. Although he did not participate in the raids directly he was seen eating meat and traveling with bandits. Herd owners who had livestock stolen from them and saw Achuka with the bandit group frequently came to Lorimet's awi demanding compensation. One time Lorimet told me that the biggest enemy he had in the world was his brother. This was just after a man took Lorimet's last milking camel in compensation following an attack by the Ngoroko. In 1984, Lorimet turned to the old men (*ekazicout*) for help. After he had presented his case, the old men in Ngisonyoka made a decree that Achuka had to leave Turkana District or he would be killed. Achuka left and by 1996 had not returned to Turkana District.

Lorimet is a very cautious herd owner who moves less frequently and subdivides his herds less often than most other herd owners. Unlike Angorot, who is often at the leading edge of group movements, Lorimet often chooses to be one of the last herd owners to move. He is thus in much less danger from enemy raids, but his livestock have less access to high-quality forage. One reason for this strategy is Lorimet's personal inclinations concerning risk; another reason is his lack of labor. Lorimet is constrained in his capacity to subdivide his herds because he does not have brothers whom he can depend

Fig. 6.4.   Lorimet

upon, and his children are not old enough to be herd managers. The one time that he made a bold decision and moved to the Naro for better grazing, while most of the Ngisonyoka remained in Nagumet, a combined force of Pokot and Karimojong raided his awi, and he lost nearly everything.

Following this disaster, Lorimet moved together with Angorot who helped by sharing food, much the same way that Angorot's uncle

Fig. 6.5.   Lorimet's wife Asekon

Fig. 6.6.   Lengess driving animals while migrating

helped him following the death of his father. Lorimet was able to gradually restock, begging livestock from friends and relatives and getting animals in bridewealth payments (for more detail on Lorimet's recovery, see chap. 10). By 1987, he had rebuilt his herds to the extent that he could separate from Angorot. In 1987 he married his second wife, although he had to defer bridewealth payments over a number of years. At the time I thought that this was an extremely poor decision. I remember thinking to myself that Lorimet seems determined to fail—just as he begins to recover he puts himself in massive debt. However, a few weeks later I was discussing herd management with a group of old men, and I asked what they thought about Lorimet's decision to marry another wife. I was astonished that there was universal agreement that this was one of the best decisions that Lorimet had ever made. They summed up their opinions concisely: "Livestock follow children." In other words, if you do not have children you will lose your livestock, and you will not be able to rebuild the herd. The decision to either transfer animals or put oneself in debt in order to bring in another wife to the family is one of the most important decisions that herd owners make. Of course, I knew that marrying for a second time was an important event, but it was Lorimet's decision to marry and the favorable opinion of the old men about this decision that made me realize just how tightly articulated family formation and growth are to herd management. I will return to this topic in more detail in chapter 9.

Atot

Atot was a gentle, likable man who was in his late forties or early fifties when I first met him. He was the youngest of three brothers, the son of the first wife of his father. His father was very wealthy when he died around 1950. Aparo, the first son, assumed the role of household head, and Atot and his elder brother, Loceyto, split off from Aparo and lived together for about six years. Soon after the birth of his first son, Atot separated from Loceyto and began life as an independent herd owner. Like Angorot, Atot's foundation herd was small: about five camels, five cattle, and a small fence of small stock. His herds gradually increased primarily through bridewealth payments and inheritance. Atot inherited a large number of animals following the death of Aparo in 1970.

Like Lorimet, Atot has had problems securing adequate labor for herding, even though his awi appears large. This need has made him to some extent dependent upon his brother, Loceyto. Loceyto allowed one of his sons, Epakan, to be brought up in Atot's awi and has assumed the role of principal manager/herder for Atot's cattle. As Atot's cattle herd grew he put some cattle in the care of another one of Loceyto's sons, Naperit. During times when the nonmilking small stock were separated from the awi, Atot often had to enlist the help of his daughters for herding.

When I first met Atot he was living with four wives (one of whom was inherited), two half sisters, and their children. His first wife, Nachok, had four sons (ages infant to twenty-one) and three daughters (ages five to twenty) living with her. Her eldest son, Lori, was the principal herder for the awi's livestock, but all the children, including the daughters, had to help with herding whenever needed.

Atot's second wife, Emuria, lived with her two sons (ages one and five) and her eight-year-old daughter. Atot's third wife, Ekrimet, lived in a third ekol along with her infant son. Atot's inherited wife Acure and dependent women (both half sisters) lived in the last three ekols. Acure lived with her three children, two daughters age one and seven and a son age thirteen. The two dependent women were both widows. One, Namentot, had married a very poor man and was left destitute when he died. She lived with her young daughter. The other widow married a wealthy man but he was sterile. When she became pregnant by another man the husband divorced her and she moved to Atot's awi. She lived with her three sons and three daughters, age one to twenty.

Although Atot's awi was large, there were only three young adults who could take responsibility for herding, and two of them were young women. Without the help of Loceyto's sons, Atot would have been in great difficulty, but even so, Atot could not separate his livestock as much as he wished, and he had to occasionally rely on his eldest daughter to manage livestock away from the awi.

In 1980, his livestock holdings consisted of 79 camels, 104 cattle, 205 goats, and 70 sheep. During the 1980–81 dry season many of Atot's cattle died of a combination of nutritional stress and contagious bovine pleuropneumonia. Atot also lost a number of camels to preda-

Fig. 6.7.   Atot and son Lori branding a camel

tors during that dry season. By the wet season of 1982, Atot's livestock
holdings had fallen to 37 camels, 36 cattle, 163 goats, and 41 sheep.

During the mid-1980s Atot moved to an area called Riet located
north of Lokori and on the southeastern side of the Kerio River. He
adjusted his migratory orbit to this new environment and rarely
returned to the areas used in the early 1980s. This was a fairly difficult
place to reach by vehicle (impossible at many times of the year), and I

Fig. 6.8.   Ekrimet milking a cow

was unable to keep in close contact with Atot's family after that. I last visited Atot in Riet in 1987. By that time his eldest daughter, Aka, had married and left the awi. His livestock holdings were 56 camels, 44 cattle, and 48 small stock. He was still having some trouble with labor even though his sons were old enough to be good herders.

Problems with herding led to a somewhat embarrassing situation in the 1987 dry season for Atot. His eldest son, Lore, had returned to the awi with the cattle, but a number of the milking cattle were missing. Atot was furious with Lore, and both Lore and Atot spent much of the night searching for the lost cattle. They returned around midnight with all of the missing livestock, but the next day Atot said that he would herd the cattle himself that day. It is very unusual for a herd owner with sons old enough to herd to go out with livestock and remain with them all day. This was meant to embarrass Lore, but Atot fell asleep under a tree and the whole herd wandered away. Atot returned alone that evening, and almost all the members of the awi had to spend the night again searching for lost cattle. All the cattle were found and no damage was done, except to Atot's pride.

Lopericho

Lopericho was a man in his late thirties when I first met him. He was full of energy and humor all the time that we worked together. When Jan Wienpahl and I were on the walking safari, people at his awi were the only ones to offer us anything, in this case tea. This small act endeared the people at Lopericho's awi to me, as well as Lopericho himself when I met him a few weeks later. My relationship with Lopericho and his family was further strengthened in 1983 when his wife named their daughter after me. I had driven Lodio to the hospital in Lokichar a few days before as she was due to give birth but was having trouble with the pregnancy. She delivered successfully in the hospital, and I was told later that there was a discussion (I do not know how serious) about whether to name the child Terry or Suzuki (after the vehicle I used to take Lodio to the hospital). In any event, I gave Terry her first beads and am considered a father (fictive) in the kinship system.

Lopericho was the first son of his father's first wife. His father was a fairly wealthy man, and upon his death in the late 1960s, Lopericho inherited a large herd of livestock. This consisted of about 100 camels, 10 cattle, and 100 to 150 small stock. Lopericho was able to build upon the herd he inherited, and in 1980 he had 80 camels, 122 cattle, and about 100 small stock. As his herds increased he transferred animals to the wives of his father, demonstrating his generosity and concern for those in his extended family.

In 1980 Lopericho lived with his wife Lodio and their children, along with the families of Nawar, his sister; Nakucho, the wife of a half brother; and the family of Nawiangorot, the fifth wife of his deceased father. In 1980, Lodio had three sons, ages two to seven. Nawar had four sons and three daughters. The boys were ages one to nineteen, and the girls were ages three to eleven. Nakucho had three small sons living with her. Nawiangorot lived with two daughters, ages thirteen and fifteen. Two older sons had left Turkana District and were working as herders on farms outside the highland town of Kitale.

Lopericho grew up herding camels and concentrated his herding operation on the raising of these animals. He once told me that he knew everything about camels, when they were sick, when they were

Fig. 6.9.   Lopericho and family

well, what they wanted to eat, and when they were happy; but with
the other animals, "I just push them around." This illustrates one
important aspect of Turkana herd management—some herd owners
tend to concentrate on one species of livestock, usually the species
they herded as boys. This was certainly true for both Lopericho and
Angorot. Lopericho, like Angorot, could utilize the labor of his full
brothers in his management strategies. His youngest brother, Iboko,
was herd manager for Lopericho's small stock. When I first started
working with Lopericho, the small stock were separated and located
far to the north around Lake Turkana. They had become isolated
there, due to drought and bandits, and I did not meet Iboko until the
following year. Lopericho kept his cattle separated into two herds.
The second youngest brother, Komosiong, managed one herd, and
Apuu, the brother closest to Lopericho in age, managed the other
herd.

Lopericho lost fewer animals during the drought of 1980–81 than
did any of the other three herd owners. His concentration on the raising of camels certainly contributed to this success, as did his friendli-
ness with young men who were known as raiders. Although he him-

Fig. 6.10.   Lodio whisking blood

self had given up raiding many years before, Lopericho often offered hospitality to men moving through the area to raid Pokot or Samburu. If a raid was successful, these individuals would often repay Lopericho for his friendship with an animal on the return journey.

Lopericho married for a second time in 1984, but he fell ill and died that same year. We (STEP researchers) felt that the cause of death was probably cerebral malaria because Lopericho was ill for only a couple of days before he died, but we will never know for sure. For a couple of years most of the family stayed together with Apol as the herd owner, but Apol did not care for Lopericho's wives well, and most of the livestock were lost in a series of Pokot raids. By the late 1980s Lopericho's family was dispersed and for the most part destitute. I will return to these sad events in chapter 9.

CHAPTER 7

Migration and Decision Making

The Four Families

This chapter describes the migration patterns and the decisions concerning mobility and herd management made by Angorot, Lorimet, Lopericho, and Atot. I will also include information concerning the climate, the social and political events that affected people's lives, and how these events influenced the decisions that were made. My goal is to provide a nuanced view of what the four herd owners were confronting each time they made a decision to move or to divide the livestock. Many accounts that examine mobility and decision making among nomadic peoples tend to isolate a single variable as the reason that a family or group moves. I have done so in the past, and in the analysis chapters that follow I identify what each herd owner told me was the most important reason for leaving one place and moving to another.

In reality, however, there are multiple concerns influencing each decision by a herd owner to move or divide his livestock. One may take precedence but all are considered. A herd owner may say that the reason that he moved from place A to place B was that the forage was better in B than in A, but there may be many places where the forage was better. Other considerations may include safety—from raiders, bandits, or predators; the locations of watering points for each of the livestock species as well as for people; the presence or absence of other families in the area; and where friends or relatives are located. Each herd owner also has an individual strategy or style in moving his family and livestock. For example, Angorot was aggressive, often on the leading edge of families moving south toward areas where raiding was common; on the other hand, Lorimet practiced a more defensive strategy, usually remaining behind the families and herds moving south, letting other families act as a "shield"[1] in the event of raids. To some extent Lorimet was able to do this because of his relatively small livestock holdings, but fear of being raided exerted a much stronger influence on Lorimet than it did for Angorot.

Although I have data for ten continuous years for Angorot and Lorimet, and two to three years for Atot and Lopericho, I include detailed accounts for only a few seasons or parts of seasons for particular herd owners; other information is summarized and condensed. I have selected periods that illustrate particular themes emphasized in the theoretical framework of the book: the variable nature of nonequilibrial ecosystems and the impact of drought; the importance of politics and history in understanding raiding; and individual variation in responding to the same climatic or political circumstances.

I start with the drought called *Lochuu*, as nearly every issue discussed in this book influenced how individuals moved and made decisions during this time. It is also a period when individual differences in responding to the same set of events can be seen in sharp contrast. Although Lochuu occurred more than twenty years ago, a very similar set of environmental and political events came together during the early part of the 1990s and again in the first two years of the twenty-first century.

January 1980 to May 1981: Lochuu

During the period referred to as Lochuu, climatic events and political circumstances combined to cause one of the most stressful drought episodes ever remembered by the herd owners with whom I worked. I began work with the Ngisonyoka in 1980, but Lochuu began in July 1979. The year was called *ekaru GSU* because the elite fighting division of the Kenyan army, the General Service Unit, conducted an "operation" in southern Turkana District with the intention of disarming the

Turkana. The operation was for the most part successful, and the Ngisonyoka were left with only a few very old British Enfield rifles. The operation did not extend south to the Pokot who were quite heavily armed with both rifles and machine guns at the time. The balance in weaponry that had existed between the southern Turkana sections and the Pokot during the 1970s was thus destroyed, leaving people like the Ngisonyoka fairly defenseless, if the Pokot chose to attack.

The period from July 1979 until March 1980 was dry, but not excessively so. During March, April, and May 1980, rainfall measured 160 millimeters, certainly enough to cause a pulse in primary productivity, but not really enough to cause a major growth spurt in the herbaceous vegetation. I met the families of Angorot, Lorimet, Lopericho, and Atot in June and July 1980, and although these families were in their wet season areas, there was very little grass available for cattle to eat. I had no way of knowing then just how serious the failure of the rains was, and the next nine months were extremely dry. By the end of this period, the Ngisonyoka had experienced almost two full years of drought, and even the camels, the most drought-resilient of the livestock species, were emaciated.

The dry season of 1979 was called *Ngikosowa*, the Turkana name for buffaloes, because during this period buffaloes around Amaler, Lochakula, and Naroo died for lack of forage. The drought called *Lochuu* (which means cool) refers to the fact that when an animal was skinned, the body was covered with "water" that was cool. People said that there was no taste to the meat or to the bone marrow. The very brief wet season was called *Lokiyo*, meaning tears. This refers to the abundant secretions coming from the eyes of diseased cattle, most likely due to contagious bovine pleuropneumonia.

Migration Patterns

The next section provides a detailed description of Angorot's migration, followed by a summary of the migration routes and decisions made by Lorimet, Atot, and Lopericho, in order to suggest the complexities involved in each choice: whether to move or not to move; whether to divide the livestock into satellite herds or keep them together; and if satellite herds or flocks are split off from the awi, who will manage them and who will provide the necessary labor. In the description for Angorot's moves I have keyed the place-names to the numbers included in map 7.1. In the rest of the chapter I have only provided the general area that people were moving (see chap. 4, map 4.2 describing the Turkana conceptualization of the landscape).

Angorot's Migration Pattern

In January 1980, Angorot's camels and small stock were far to the south of Ngisonyoka territory, and the cattle were in a high pass between Lotaruk and Kailongkol Mountains. As mentioned earlier, it was dry, but this was not the primary concern of Angorot or his brothers, who were managing the satellite herds. In October 1979, the GSU caught Angorot's brother, Aki, with a rifle. He was arrested and sent to prison for five years. Aki was the principal herder of the cattle, and his absence resulted in a serious management problem. Following Aki's arrest, Angorot decided to place his brother Erionga in charge of the cattle, but he was inexperienced, and his herding and decision-making abilities were a major concern. Angorot's awi and livestock were dangerously close to the border with the Pokot, and because most of the Ngisonyoka were now without firearms, they were more vulnerable to attack than at other times.

In January 1980, Angorot's awi was located at map point (1), at Loesekon, named for the esekon trees (*Salvadora persica*) that grow along the banks of dry wash. The leaves of these trees were important forage for the small stock, and there was water available in nearby water holes. Toward the end of January, Angorot moved south to Namerikabyekuny (2) and was joined there by Erionga and the cattle herd. Here, an adakar was forming with Turkana herd owners living in close proximity for protection against Pokot raids. Although this added to the security of people living in the adakar, it put extreme pressure on the local forage resources.

The entire adakar was forced to move a short distance to the south to seek better forage and remained at Akujemuth (3) for about a month. Although people were apprehensive about being so close to the Pokot, no attacks materialized. In fact, some Pokot and Turkana exchanged small stock and tobacco, and in March, Angorot moved away from the adakar to get better forage at Napawoi (4).

It began to rain in late March, and the rain continued through May. When Angorot felt that there was enough forage for his livestock, he began his return journey back to his home area in the Toma. The cattle also began a northward migration, but on a different route than the

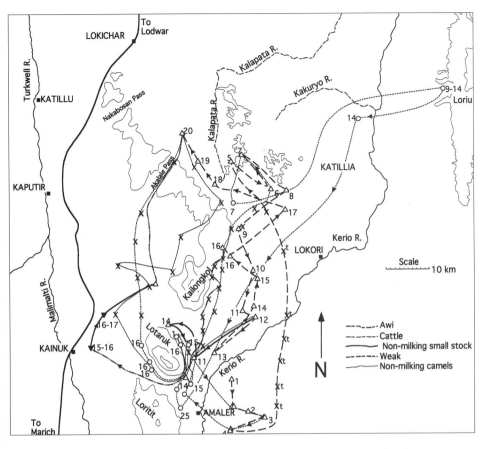

Map 7.1. Angorot's movements, January 1980–May 1981 (Lochuu)

one used by the major awi. Angorot spent most of the month of April moving north, stopping for a few days at places that he had used on the trek south. Instead of stopping at Kadapadwell, his usual wet season area, he moved north to Katamanak (5), at the base of the lava-strewn Kaweri-weri Plateau. He remained here for about a month and then moved to the sandy plains of Edome (6).

When the awi began to move north, the cattle moved to the slopes of Loretit Mountain, waiting for the grass to begin to grow in Nadikam and the Toma. The cattle moved in mid-May and joined Angorot's awi while they were in the Toma, at Edome.

By early June all the livestock and family members were together. This is normally the best time of the year for the Ngisonyoka. Typically there is abundant forage in the Toma, and there is enough vegetal diversity for all the livestock to have access to good forage. In non-drought years there are usually pools of standing water, and this makes watering the livestock very easy. When there is enough moonlight to see, young people frequently dance and sing all night long. Some of my most striking memories from the early days of fieldwork involve lying awake at night hearing the singing in the distance and the tinkling of bells as the livestock shook their heads in the light breeze.

However, by late June 1980, people were becoming concerned with the lack of forage in the plains. There was some grass, but the cover was very patchy, which made herding the cattle a very difficult task. Angorot's awi moved three times from June to mid-August while remaining in the Central Plains (7–9). The cattle moved away from the awi in early July and then returned for ten days at the end of the month. In early August, the cattle moved to the east on their way to the Loriu Plateau in search of grass.

In mid-August Angorot began to move his awi southward to the gravel plains of Nadikam (10). Many Ngisonyoka were also moving south, but not in an adakar; each family moved separately. By mid-October Angorot's awi was located close to his wells at Kaakolit. Forage was sparse here, and for the first time, I saw herders climb trees to cut branches so that baby goats could eat the leaves. As more people congregated along the water-bearing stretches of the Kaakolit River, Angorot decided to move his awi about two kilometers away from the wells to reduce the pressure on the forage close by. He also decided to separate his milking and nonmilking small stock at this time. The non-milking small stock, under the direction of Angorot's brother Nakwawi, moved to the south, closer to the foot slopes of the central mountains. Here they had better access to forage but were much farther away from water than they had been when they were at the awi.

About three weeks after the nonmilking small stock separated from the awi, Angorot also separated his nonmilking camels. Like Nakwawi with the nonmilking small stock, Akopenyon took the non-milking camels to the south and closer to the mountains. However, he had to be careful with the camels as they often stumble and fall on rock-strewn and broken ground, occasionally breaking their legs.

Although no raids had occurred in the vicinity of the awi, stories of Pokot raids in other areas were common, and people were nervous. The intertribal boundary between the Turkana and the Pokot bisected Loretit Mountain, with the northern slopes belonging to the Ngisonyoka and the southern slopes belonging to the Pokot. The drought that brought the Ngisonyoka to the mountains to the south also brought the Pokot to the mountains in the northern part of their territory. In addition, the emeron Kalinyang was having dreams predicting Pokot attacks in the immediate future.

Toward the middle of November Angorot moved the awi south to a dry wash called Komykuny (13). Jan Wienpahl and I were living and traveling with Angorot's awi during his stay at Kaakolit and on the move to Komykuny. As we moved farther south, tensions began to mount, and fear of Pokot raids intensified. On November 17, Angorot built his awi beneath a large rock outcropping, frequented at dawn and dusk by a troop of baboons. The people in the awi were disconcerted to hear the baboons barking at night as they said this happened when people were moving through the bush. In their minds this meant raiders were near. This time people's fears of a Pokot attack were realized.

November 22, 1980

On November 22, about an hour before sunset on a night of a full moon, Jan and I were helping butcher a goat when we heard children running up the Komykuny wash yelling "Upe, Upe" (Pokot, Pokot). Everyone jumped up and began questioning the children, who said that a group of Pokot had just raided Angorot's small stock that were returning from getting water in the Kerio River. The raid occurred on the other side of a lava hill, about a half-kilometer away. Some women began running into the bush, afraid that the Pokot raiding party would soon attack the awi. Because no one at Angorot's awi had any guns, engaging the attackers entailed considerable risk. We soon learned that the entire flock of milking goats and sheep had been stolen, and the two herders, boys ages nine and thirteen years old, had been killed.

Angorot came up to us and suggested that we take some men and drive to Amaler, a settlement about forty kilometers south where there was an outpost of the Kenyan army, to report the raid. The remaining men would track the Pokot and see if they could get back some of the stolen animals. As Jan and I began driving south, night fell and a heavy rain began. In the vehicle we were tense, wondering if this was a small raiding party or a large force that we might encounter on the trip to Amaler. When we reached the army post, the Turkana men reported the raid and the commander asked us to carry some soldiers back with us. About fifteen soldiers got in the back of our pickup and we headed back to the awi. When we were halfway back the soldiers asked us to stop, and eight or nine of them got down from the truck and took off into the bush with most of the other Turkana men. Our interpreter, Elliud, said that they were going to a point on the Kerio River where they expected the raiders to try to cross and were going to set up an ambush. We proceeded back to the awi.

There we found a nearly abandoned camp, and most of the women had hidden in the bush. The few people that were around told us that all the men had followed the Pokot and that the remaining people were afraid of another attack since all the men were gone. Elliud suggested that we drive to Lokori and report the attack to the councillor of the area, a friend of both Elliud and Angorot. So once again, we took off in the dark to report the raid; by this time the rain had stopped. Lokori was about an hour and a half drive, and we arrived tense and tired. After talking to the councillor we again returned to the awi. At the time we believed by returning and staying the night, we offered a degree of protection to the people in the awi. We thought that Pokot raiders would be reluctant to attack the awi while we were there.

The night was filled with the bleating of young goats and sheep calling to their mothers and needing milk. In the morning, people returned and milked the camels. Most of this milk was fed to the kids and lambs as people in the awi tried to keep the young animals alive. This went on for two more days, but many of the young goats and sheep died, despite these efforts.

The soldiers and the men from Angorot's awi caught up with the Pokot raiders the next morning after the raid. A battle was fought south of the ford where the ambush was set, and the Pokot fled into the bush, abandoning the stolen livestock. Although the raiders were pursued for another day, they were not encountered again. While some of the men chased the Pokot, the others rounded up as many animals as they could find. In all, 200 goats and sheep were recovered, and 151 lost. About three-quarters of the kids and lambs also died.

One of the boys who was killed, Tioko, was the son of Angorot's father's third wife, Agule, and thus Angorot's half brother. The other herder killed was a relative living with Angorot's mother and helping with the small stock.

There are times during fieldwork when breakthroughs happen in the relationship between local people and researchers. Although this raid was disastrous for Angorot's family and caused Jan and us to seriously reconsider South Turkana as a viable field site, the fact that we had experienced this raid together with Angorot and his family resulted in a bond that endures today. When visiting Angorot in 1996 he told me that I should consider his animals as my animals and reminded me of the fact that he could have lost all his small stock without my help in 1980. Although this is certainly an exaggeration, ever since that raid I have felt particularly close to Angorot and all the members of his family. I believe that this also changed how other Ngisonyoka perceived us. We became more a part of the community and were considered less as outsiders.

Two days after Angorot returned with the remaining small stock the awi moved, continuing the southward migration. However, after only five days at Kabanyiet, Angorot moved again. The high density of ticks in the area was one factor, but the continued tension and fear of further Pokot raids influenced Angorot to reverse his direction and he moved north—away from the Pokot.

Angorot again located his awi along the banks of Kaakolit wash, where he remained for most of December (14). Forage was already difficult to obtain here, and on January 1, 1981, Angorot moved further north to the foot slopes of Kailongkol Mountain (15). Here people again had to cut branches of trees so that the young animals could have enough to eat. Although this area was relatively safe from Pokot raids, Turkana bandits, known as Ngoroko, had left the Loriu Plateau and were moving south close to where Angorot had his awi. These

Turkana raiders were ostensibly moving south to attack Pokot or Samburu, but they had recently begun to steal animals from Ngisonyoka, as well. As was mentioned in the discussion of raiding (chap. 5), it is considered a common courtesy to kill an animal for meat when Turkana raiders pass through, knowing that hospitality would be rewarded with a gift of livestock on the return from a successful raid. However, raiders from the north began taking livestock by force, and by the early part of 1981, reports of stolen milking animals, rapes, and even killings began to be attributed to Ngoroko. To move away from "the road of the Ngoroko," Angorot moved his awi further north and west (16). This was a difficult time as grazing was poor and most of the small stock and camels had to travel twelve kilometers each way to water at the Lokwamosing spring. The stronger of the two flocks of the nonmilking small stock joined the awi there.

In mid-March the rains broke, and Angorot saw that there had been a few small showers in the area around the small isolated volcanic mountain called Kadapadwell. This is Angorot's home area (*ere*), and he moved there on March 24, but the grass had yet to grow and the other vegetation was just beginning to turn green. After remaining for only ten days he moved northward again to the base of the Kaweri-weri Plateau (18). After two weeks there he moved westward where a large adakar was forming. The adakar moved south in mid-May and to Atot a Lomogkerion (20), and Angorot's satellite herds joined the major awi.

Following their separation from the major awi in January, the nonmilking camels moved to the southern flank of Lotaruk Mountain. From here they moved to the north side of Loritit Mountain, but this was both far from water and close to the Pokot. They moved back to the south side of Lotaruk and remained there until the rains broke in mid-March. Just before beginning their northern migration back to the Toma, Ngoroko stole three camels, prompting Akopenyon to leave the dangerous southern territory as soon as possible.

After leaving the awi in mid-October, the nonmilking small stock found forage on the southern flank of Kailongkol Mountain. In late November, Nakwawi moved the goats and sheep to a high pass between Lotaruk and Kailongkol. Forage was sparse in the highlands, and Nakwawi moved to lower elevations searching for greener vegetation. Unfortunately, there were many ticks in the lower elevations, and the small stock moved again, this time to the southern flank of

Lotaruk. In mid-January Angorot thought that the small stock were in trouble because of the very sparse forage. He made the decision to split his nonmilking small stock into two flocks: one flock composed of relatively strong animals and one flock composed of animals that Angorot considered so weak that would probably die without better forage than they were now getting. The stronger flock was to move north, away from the danger of Pokot raids and closer to Angorot's main awi. The weaker animals were to move west to the banks of the Malimalti River. This area was known to be infested with ticks and mosquitoes, and the chances of serious livestock disease were high. However, there was green vegetation there, and Angorot surmised that if the weak small stock did not get better forage they would die anyway.

These animals remained in the Malimalti region until the rains began and then moved to Lobokot to join other Ngisonyoka waiting for the grass to grow in the Toma. In what seemed like a bizarre twist of fate, the coming of the rains did not spell relief as expected. The first rains came at night, and I was in Lobokot at this time. I visited many camps of Ngisonyoka small stock the next day and was shocked to see hundreds of animals, mostly sheep, lying dead inside their thorn fences. What had happened was that the already severely stressed goats and sheep had become wet and chilled in the night and began to shiver. Shivering increases the metabolism in order to produce heat, but it also uses energy, more energy than many of the small stock had stored. According to the ecologists with me at the time, the animals literally shivered to death.

Angorot's cattle, under the management of Erionga, had left the Toma plains in late July 1980 and moved to the highlands of the Loriu Plateau, where they remained until the end of November. The Loriu Plateau is located in Ngiesitou territory, but due to a long-standing relationship between the Ngisonyoka and Ngiesitou, no permission is required to take cattle up onto the plateau. During this time, the Loriu was the staging ground for numerous Ngoroko raids against the Pokot and Samburu. The raiders were also stealing cattle from the local herders, and Angorot was hearing reports of Erionga losing cattle both to bandits and lions. He was also concerned about rumors that Erionga was killing many cattle for food to impress his friends who were also herding cattle on the Loriu.

In late November, Angorot sent word to Erionga to move the cattle off the Loriu and to bring them down to Loritit Mountain where the danger from Ngoroko was reduced, but the danger of having the cattle taken by Pokot increased. Still, here Angorot could at least keep better track of what was happening with the cattle herd. In February 1981, the cattle moved to the northern slopes of Lotaruk, and then in early March they moved to Lotaruk's southern slopes. Here there was forage, but the cattle were a full eight to nine hours walk from water each way. In addition to the stress caused by long walks to water and inadequate forage, there was an outbreak of contagious bovine pleuropneumonia. During this time, twenty-three cattle died from CBPP, and ten more starved to death. In mid-March the cattle moved into the Western Plains (Naroo) and watered in the Malimalti River. When the rains broke, the cattle remained in Naroo and slowly began the journey back to the Toma along with many other Ngisonyoka cattle herds, arriving at Angorot's awi in late May.

Lochuu was disastrous for all of Angorot's livestock. He lost a total of 73 cattle, which represented 69 percent of his total cattle herd. Twenty-three animals died from contagious bovine pleuropneumonia, 14 others died from starvation, 10 were stolen by Pokot, and 8 more were stolen by Ngoroko. He also slaughtered 6 head of cattle for food during the dry season, and 12 others died from a variety of other causes. The drought was almost as bad for Angorot's camels. Fourteen camels died from starvation, and 5 others died from disease. In addition, bandits stole 5 camels, hyenas killed 2, and lions killed 2 others. A total of 34 camels died, were killed, or were stolen during the drought; this was 64 percent of Angorot's total camel herd. The year was equally bad for Angorot's small stock. He entered the 1980 dry season with a total of 722 goats and 269 sheep; by the 1981 wet season he had 343 goats and 62 sheep. Thus 586 small stock died, were stolen, were slaughtered, or were sold during the drought. The total losses for small stock represented 59 percent of Angorot's holdings in goats and sheep.

Lorimet's, Lopericho's, and Atot's Migratory Patterns

Lopericho, like Angorot, was fortunate in having a number of brothers who were able to help with the management of satellite herds. Although Atot had one teenage son, he relied on his older brother Loceyto to help with the herding of his cattle. The labor problems that Atot would eventually experience were recognized soon after he was married, and Loceyto "lent" one of his young sons to be brought up in Atot's awi. The boy, Epakan, eventually became the primary cattle manager for Atot and was responsible for a mixed herd of Atot's and Loceyto's cattle. Atot also kept some cattle with Naperit, another one of Loceyto's sons. Thus, although Atot did not have enough labor to divide his livestock, these problems were able to be overcome.

Lorimet, on the other hand, had serious labor and management problems, even though he had two adult brothers who should have been a great help to him. One brother, Lengess, left Turkana district in the mid-1970s following a Pokot raid in which he was shot in the leg. He decided to try to find work on a ranch outside the highland town of Kitale. Although this work paid very little money and the whole experience was demeaning, he at least felt safe. Lorimet's other brother, Achuka, turned out to be a very irresponsible young man, sometimes leaving a herd that he was managing in the care of others while he went into town to drink and visit girlfriends. It came as no surprise that the livestock under his care suffered much higher losses than did similar herds. He eventually joined with a group of Ngoroko, and although he did not hurt anyone or personally steal livestock, he was banished from Turkana district by the elders, or *ekazicouts*, of the Ngisonyoka.

When I first met Lorimet in 1980, his cattle were being managed by the sons of his father's brother. Unfortunately for Lorimet, this did not turn out to be a very satisfactory arrangement. Lorimet felt that many of the cattle deaths reports by his cousins were actually animals that had been slaughtered for food or lost due to lack of proper care.

When I met Lorimet, Lopericho, and Atot, their major awis were located in the Toma, near the small mountain called Kadapadwell. Lorimet and Atot had all their small stock and camels with them, but the herds and flocks owned by Lopericho were spread out all over Ngisonyoka territory. During the previous dry season all three herd owners were living in the southern part of Ngisonyoka territory, and each had subdivided their livestock into species-specific herds. For many of the herds and flocks, movement was a response to both environmental conditions and the presence of patrols by the Kenyan army. As the dry season progressed each herd owner migrated further south, while the cattle herds moved higher into the mountains. When the rains broke the major awis moved back north into the Toma. Some of the herd owners and their satellite herds made specific moves in order to leave an area where the Kenyan army was patrolling. Most Turkana viewed the Kenyan army in the same way they viewed an enemy raiding party. Livestock would be confiscated (stolen as far as the Turkana were concerned), and people could be beaten up or killed and women raped.

The most unusual migratory pattern was that of Lopericho's non-milking small stock managed by his younger brother Iboko. Iboko had crossed to the east side of the Kerio River during the dry season and could not cross back once it started to rain. Iboko followed the eastern bank of the Kerio on his migration back north but realized that a large group of Ngoroko were only a short distance behind and taking the same northern route. He was able to cross the Kerio at the bridge in Lokori, but instead of going northwest into the Toma, he moved into the dissected gravel hills in the northeastern part of Ngisonyoka territory. When I asked him why he did this he said that he thought that he needed to get as far away as possible from the Ngoroko and that he thought that they might be moving toward the Toma if they were able to cross the river.

Following the abbreviated rains, Lorimet, Lopericho, and Atot left the Toma and began moving south along the same route taken by Angorot. All the cattle herds remained separate with some herds moving to the Loriu Plateau while others migrated south to Kailongol Mountain. By November 1980 all the cattle, with the exception of the herd managed by Lopericho's brother Apuu, had moved to the southern mountains Lateruk and Loritit and remained there for the rest of the dry season. Apuu migrated all the way across Ngisonyoka territory and crossed the Turkwell River where he remained during the dry season, moving along the banks until the rains broke.

As herd owners moved south they each also began to divide the livestock. Lorimet separated his camels from the awi and put them in the care of his brother Achuka. However, unlike almost all other Ngisonyoka herd owners, Lorimet decided that the continued migration south posed too much of a risk from Pokot attacks, and in November he decided to turn back north. He moved into recently abandoned awis for a few weeks at a time and continued northward into the Toma. He could do this because he had relatively small livestock holdings, and there was no competition for grazing from other herds. He also moved far less than other herd owners during this period, choosing to have the livestock conserve energy as the forage resources were drying out. By mid–dry season Achuka was tired of managing the camels and left them in the care of Angorot's brother Akopenyon, while he went into the town of Kainuk to be with a girl-friend. When the rains broke and the camels returned to the Toma, Achuka was still in town, exacerbating Lorimet's already difficult labor problems.

Atot and Lopericho continued south along with Angorot but divided their livestock in a way not seen again during my sixteen years of research with these families. During the middle part of the dry season awis were being attacked by both Pokot and Ngoroko. Although the Ngoroko were not killing people, they were raping girls and taking fat and pregnant camels, something they had not done before, signifying a shift in their raiding tactics. In order to protect his daughters and pregnant camels, Atot separated them and had his eldest daughter move them all the way to the Naroo. He also sepa-rated his small stock under the care of Lorei, but this caused some sig-nificant labor problems for Atot who had to herd the remaining camels himself. Lopericho took all his livestock and built a separate camp for them about two kilometers away from the awi. Each morn-ing the women would get up and walk down to the camp to milk the camels and goats and return to the awi to feed the family. Although the livestock remained in danger from raids, the people were pro-tected by being separated from the herds.

Like Angorot, when the rains broke Lopericho moved to the Naroo and then north to the Toma. Fearing Ngoroko attacks, Iboko, along with the small stock, continued to move north. He spent the entire dry season along the banks of Lake Turkana, returning to the Toma in early May 1981.

Atot's cattle herds, managed by Epakan and Naperit, also moved to the Naroo, following the onset of the rains, but Atot migrated back to the Toma following the same route as his migration south. Lorei accompanied the cattle to the Naroo and on to the Toma

The rains broke the drought that had lasted for almost two years, and even though each herd owner responded differently to the com-bined perturbations of drought, raiding, and the threat of violence, each lost a significant percentage of his livestock. The serious out-break of contagious bovine pleuropneumonia that struck Angorot's herds also infected Lorimet's cattle. While located on the slopes of Lotaruk, 25 of Lorimet's herd of 45 died from CBPP, and although no other cattle died, this represented a 54 percent decrease in his cattle herd. He lost 19 camels during Lochuu; Ngoroko stole 10 camels. Predation by hyenas and the slaughter of camels for food were the other major contributors to the 63 percent decrease in his camel herd. According to Lorimet, "many" of his small stock died from starvation during the drought, and my records show a decrease from 268 animals to 155 small stock, a loss of 42 percent.

Despite all the attempts to secure forage for the livestock herds and to avoid attack by enemies, Lopericho still suffered significant losses in all his herds. His strategy for avoiding Pokot and bandits worked well, but his camels were still quite stressed nutritionally, and he lost 15 camels to starvation during Lochuu. He also lost 14 camels due to disease and slaughtered 3 for food. Another camel was lost and not recovered, and one fell and broke its leg and had to be slaughtered. In spite of the drought year, Lopericho gave 3 camels away to help needy relatives. He began the drought with 80 camels and by the time the rains broke he had lost 37; this represented a 43 percent loss.

The cattle did not fare any better, even though both Apuu and Komosiong prevented any losses to Ngoroko or Pokot. The major problem affecting the cattle herds was an outbreak of contagious bovine pleuropneumonia, which struck in late November and December. Fifty-eight animals in total were lost, with Komosiong's herds suffering more losses than Apuu's. Thus Lopericho emerged from the drought with 52 percent of the cattle that he had when the drought began.

I did not have a chance to count Lopericho's small stock before they moved to the shores of Lake Turkana, so I cannot calculate exactly how many were lost. Iboko told me when he rejoined the awi that "many" sheep and goats had died from starvation. He only returned with 26 animals, and I suspect that the flock that returned contained less than half the animals that began the arduous northern migration.

Atot's cattle were also devastated by the outbreak of contagious bovine pleuropneumonia that swept through the southern Ngisonyoka area in the months of November, December, and January. Atot lost 142 cattle to CBPP during this time. In addition, Atot lost many others to predation by lions and had to sell 10 others to buy food. In total he lost 86 percent of his total cattle herd. He began the dry season with 318 head of cattle and by the following wet season was left with only 44. He also suffered a loss of 40 camels, which represented a 51 percent decrease in his camel herds. Although Atot lost many camels and cattle during Lochuu, his flock of goats fared quite well. During the dry season he sold many goats in order to buy maize meal, and his total holding dropped about 25 percent over the whole drought year. The same cannot be said for his sheep; many died at Lobokot, following the rain, and others died of nutritional stress throughout the dry season. Atot began the wet season in 1980 with 70 sheep, and at the end of the dry season, only 16 were left.

Although the events described here happened more than twenty years ago, they set the stage for how people moved and used the land for years to come. They also had a major influence on how I perceived the tension in the decision-making process between moving to gain access to better forage resources and moving to increase the level of security for the family and the herds.

April 1981 to July 1982: The Move North

The remaining sections of this chapter will further illustrate some of the major themes described in the initial chapters of the book. However, I will not describe the movements, decisions, or events in the detail used for Lochuu.

The wet season of 1981 was called *ekaru Loukoii,* meaning the year of contagious caprine pleuropneumonia. Over 225 millimeters of rain fell during the months of March, April, and May, with small amounts falling in June and July. The total of 225 millimeters was the most cumulative rainfall recorded for any three-month period during the entire decade. The heavy rain further weakened the already stressed small stock, and during the months of June and July many goats died of CCPP. Beginning in July, an eight-month-long dry season set in with only 21 millimeters of rain falling from the end of July to the following April. The dry season was given two names: *ekaru Kalimnarok* or *ekaru Kakong.* Both are places in the western Naroo where Ngisonyoka have deep wells, and the Pokot raided both during the 1981–82 dry season.

The entire year was unusual from the standpoint of land use for many Ngisonyoka. The adakars that formed in the northern Toma remained together for much of the year. In addition, instead of moving south as is typical for Ngisonyoka during the dry season, the large adakars located in Toma moved north, the only time that this occurred in the decades of the 1980s and 1990s. The heavy rains provided a significant pulse to the ecosystem, and many of the discussions under the tree of the men, and within the different adakars, concerned the possibility that the forage resources might be good enough to get through the dry season without moving south. The continued fear of raids permeated nearly every discussion, and tension was high throughout the wet season and as the dry season set in.

The stress felt by herd owners and their families following the drought and Pokot raids of the previous year was compounded by a dramatic increase in attacks by Ngoroko in the wet season of 1981. The danger to the people that we were living and working with was illustrated on the night of April 13.

April 13, 1981

The research team was camping next to Lorimet's awi, and we had all gone to bed early as it was beginning to rain. Around 11 P.M., Lorimet came down to my tent and woke me up saying: "Terry, Terry wake up—Ngoroko are coming." Lorimet and some of the neighboring herd owners wanted us to report what was happening to the police and return with them if we could. I began to hear automatic rifle fire and started to wake the members of the research team. As the sound of gunshots approached, two vehicles, carrying some Turkana and all members of the research team[2] except Jan and myself, left in a driving rain, pulling one of our tents down and running over a stove in the rush to escape. Jan and I did not want to leave without Lopeyon, and we could not find him. Soon we started to hear people shouting and whistles blowing; people appeared and disappeared in the beams of our flashlights running through the rain. Lopeyon had been at a neighbor's awi and returned shortly after he saw the two research vehicles heading off toward Lokori. We left shortly after that, but were unable to report the attack until the next morning as we could not raise anyone at the police post.

We returned the next day with a few policemen and learned that over twenty awis had been attacked, and most herd owners had lost some small stock and camels. The loss of livestock was bad, but it was the viciousness of the attack that shocked everyone. A baby had been thrown into a fire (but rescued before any serious harm was done), and although no one was killed, many of those who fought with the Ngoroko had been badly beaten, and a few had knife and spear wounds.

This attack changed the way that the Ngisonyoka felt about the Ngoroko and how they responded to them. Everyone recognized that the problem was becoming serious, but most felt powerless to confront the Ngoroko without guns. Two days later a large meeting was called, and it was made clear that all men, both young and old, needed to attend. The men were divided into two groups; the men whose generational set were called "stones" or "mountains" sat together, as did those referred to as "leopards." This was the first time I had seen any form or organizational feature associated with age or generation. During the course of the day one man after another rose and spoke, emphasizing his major points with thrusts of his herding stick. One of the most eloquent speakers was Angorot, who spoke softly but with passion and commitment. Angorot emphasized the need to support each other and to fight the Ngoroko. "We should share our fate together and if we are to die, then let us die together. If the Ngoroko come, now we will fight. . . . We need to defend ourselves, not ask for help from the towns of Lokori or Lokichar." Another man, Nakwan, who was also attacked, followed Angorot and berated some of the older men who were emphasizing caution. Gesturing wildly with his stick he said, "All this talk is nothing—if all we do is talk, then we might as well go move to Lokichar or Lokori." As the afternoon wore on many of the young men began to leave saying that the time for talking was over, and they were going to fight, even if they had to fight with spears against the Ngoroko. Eventually a consensus was reached whereby the "mountains" would go to the mountains and spy on the Ngoroko and the "leopards" would begin preparing for a battle against the Ngoroko.

The Ngoroko must have learned about the meeting and the resolve of the Ngisonyoka to fight against them, as they moved away the next day, far to the east and south. Although this did not end the problems caused by the Ngoroko, it did break the tension and allowed people to get back to their everyday lives. For me it revealed a level of social organization of which I had heard, but never seen. The willingness of the young men to take matters into their own hands, regardless of what the elders (*emerons*) said, emphasized to me just how individualistic Turkana society could be.

The decision to move to the north had significant drawbacks. First, the area was to a large extent unknown to most herd owners that I knew. Second, no one owned wells there, and people and livestock would have to depend on shallow wells along the Kalapatta River or the remaining pools of water in the Kerio. Third, even the wells in the upper Kalapatta were questionable, because people were unsure of the location of the sectional boundary between the Ngisonyoka and the Ngibocheros. If the Ngisonyoka crossed the boundary without permission, both the grazing and water resources could be denied to them.

Both Angorot and Lopericho moved north in December 1981, along with the large Ngisonyoka adakars (map 7.2). Atot moved to the north and east in October, but in December, instead of moving toward the Ngibocheros border with the Ngisonyoka adakars, he crossed the Kerio and spent the dry season in the gallery forest south of the town, Katilia.

Lorimet remained in the Toma with the other herd owners, but he was not convinced that the threat from the Ngoroko was over. While the majority of the men discussed the pros and cons of moving to the north, Lorimet was considering moving away from the areas that he felt were still threatened by moving across the mountains to the west, into Naroo. An adakar was forming there made up primarily of herd owners and their families who considered the Turkwell as their home area. Lorimet decided to leave his camels and remaining cattle in the Toma, but he moved with his small stock to join the adakar in the Naroo in December 1981. The day after he arrived a large raiding party of Pokot and Karimojong warriors attacked the adakar. The raiding party seemed unusually well-equipped with machine guns, rocket-propelled grenades, and possibly a flamethrower; it was astounding that no Turkana were killed in this raid, but livestock losses were very high. Lorimet lost all his small stock and returned to the Toma, depressed and on the brink of destitution. He followed the

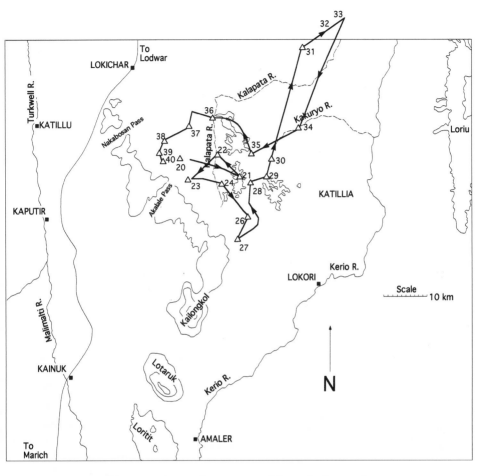

Map 7.2.   Angorot's movements, May 1981–June 1982

other herd owners north with the few camels and cattle that he had left. He joined together with Angorot who was willing to share food with Lorimet and his family. They remained together until the wet season of 1987, by which time Lorimet's livestock had increased to the point where he could again resume an independent life-style.

Angorot, Lopericho, and Lorimet remained in the northern part of Ngisonyoka territory throughout the dry season, but life was difficult for everybody. There was little shade for livestock or people, access to water was problematic, and the Ngibocheros were harassing

Ngisonyoka herders at water holes and sometimes while the herds were grazing. The rains began on April 1, 1982, and soon after that most Ngisonyoka began to migrate south.

During this period, Angorot's small stock numbers stabilized and began to increase, as did his cattle. His camel herd continued to decrease, however, primarily due to losses to Ngoroko, sales, and giving camels for food to needy friends and relatives. Lorimet was devastated by the Pokot raid and was only able to remain in the pastoral sector because of his brother-in-law's generosity. Angorot treated Lorimet and his family in much the same way as his uncle treated him and his brothers following the death of his father. Without this mutual aid ethic among the Turkana, many more families would have been forced to resort to begging in town or to settle along the Turkwell River where they would try to eke out a living as cultivators.

Lopericho's herds fared better than the other three herd owners' during this time. Both his cattle and camel herds increased following the birth of calves in the wet season. His small stock flock stabilized but had not yet begun to increase.

Atot moved back to the Toma following the onset of the rains, but his herds continued to decline, primarily from disease. The 1981 dry season marked an important transition for Atot. He decided to relocate his *ere* (home area) to Riet, where he had spent the dry season, on the eastern side of the Kerio River. He said that the forage resources of the area were good, and he got along well with the mixed Ngisetu/Ngisonyoka groups that lived in the area. He also said that the Toma had too many hyenas. Considering the number of Atot's animals that were lost on a regular basis, this probably made good sense. I also suspected that Atot was interested in moving away from the influence of his wealthy elder brother Loceyto, but he never told me this.

Dry Season 1982–83 to Dry Season 1984–85:
Stabilization and Recovery

From this point on I am going to summarize movements and decisions made by Angorot and Lorimet. Following Atot's decision to relocate his ere to Riet I was only able to visit him occasionally. Although I tried to reconstruct his migration patterns, he was moving in areas that I was not familiar with, and so I do not feel confident enough to report what influenced his decisions to move and divide his herds. Lopericho died in 1984; I will discuss what happened to his livestock and family following his death later, in chapter 9.

A two-year period of recovery and modest herd growth began during the dry season of 1982–83 and lasted until the dry season of 1984–85. Following this, another severe drought caused great hardship throughout Turkana district, although the drought was far milder in southern Turkana than in the rest of the district.

The rains stopped in July 1982, but the dry period lasted only for a few months. A small amount of rain fell in each of the months of October, November, and December. Although January 1983 was dry, rain again fell in each of the following three months. This was considered a very good time for the Ngisonyoka because there was some precipitation throughout the year and the vegetation remained green. The Ngisonyoka refer to this period as the year of the yellow sorghum, because the farmers were able to harvest sorghum three times during that one year at the irrigation schemes at Morulem (north of Lokori).

Angorot and Lorimet spent the months of July through February moving between the Toma and eastern lava plains. One of the main reasons for moving was the renewed presence of Ngoroko. A battle between the Ngoroko and Ngisonyoka home guards resulted in the death of two home-guards but convinced the Ngoroko to move on; although it was not confirmed, most people felt that they moved south to the Loriu Plateau.

In February Angorot migrated all the way to the banks of the Kerio River, near where Atot was now living. He separated his camels from the awi because he said that camels that did not grow up eating the forage that was available there would get sick. Angorot also decided to separate the nonmilking small stock, and Akopenyon left and traveled north with the camel herd, accompanied by the nonmilking goats and sheep, managed by Lokora.

Angorot left the Kerio area in March 1983 and for the next few months slowly moved north along the same route as Akopenyon and Lokora. In August the herds and flocks joined together, near the Nakwakal Plateau, but Angorot kept the cattle separate in a camp only a couple of kilometers away from the main awi. Rains fell in August, and many of the goats and sheep gave birth. Angorot remained in the same general vicinity until December but kept the cattle, nonmilking camels, and nonmilking small stock in separate but nearby camps.

This period of time, from July up until December, was considered a wet season and called *Aremere Morulem,* referring to an attack on Morulem by the Pokot. Again the Ngisonyoka felt that this was a good season for livestock, although the total amount of rain that fell was low. What made the difference was that some rain had fallen during twelve of the previous fifteen months, and although there was not a great flush of primary productivity, there was enough vegetation for the livestock to remain healthy and to reproduce.

While at Nakwakal, Akopenyon was married, and many animals were transferred in bridewealth. The fact that Akopenyon was married at this time and bridewealth transferred is an indication that Angorot felt that his herds had recovered to the point that he could give animals away and that he felt fairly secure about the future.

In January 1984 the vegetation began to dry out; Angorot referred to this time as the beginning of the dry season, but this was, in fact, the beginning of another drought. In the next eight months only 25 millimeters of rain fell. Angorot began a slow migration back to the Toma and remained there until September 1984. He kept his cattle and nonmilking livestock separate, but close, and they all made a number of short moves as conditions worsened.

By May, however, drought was gripping much of north and central Turkana, a drought every bit as bad as Lochuu. Although there was "white grass" for the livestock to eat in the southern part of Ngisonyoka territory, Angorot was beginning to get quite concerned about overgrazing as people and livestock from Ngibocheros moved into Ngisonyoka territory.

The Ngibocheros had requested permission to enter Ngisonyoka territory, and this was granted by the powerful emeron Kalinyang. This season is often referred to as the year that the Ngibocheros moved to this area. It is also referred to as *ekaru ngimesekin engora,* meaning the year of the brown sheep. It was called this because Kalinyang sacrificed two brown sheep in the hope that the rains would come.

As the forage dried and became more patchy, Angorot started to move south. The Ngibocheros also continued to move south, and in the beginning of October they were attacked by the Pokot. Angorot was concerned by the viciousness of the Pokot attack, although he was not personally affected, and reversed direction moving north, back to the Toma, leaving the nonmilking small stock behind.

There was hardly any grass in the Toma, and Angorot sent the cattle, managed by Erionga, far to the very rugged mountainous area south of the Kerio River. This area is known to have grass when other areas do not, but it is also dangerously close to the Pokot.

Erionga had sent word that the grazing conditions were good near the abandoned town of Napeitom, and that the cattle were doing well. In January 1985, Angorot felt he had no choice but to move south, close to the cattle herd. While in the Napeitom area, Angorot kept the camels and the milking small stock with him, while the nonmilking goats and sheep and the cattle went off on their own. In March, the Pokot attacked 57 awis and satellite camps. Angorot's cattle camp was among those attacked, and he lost 110 cattle, leaving him with only 20 to 30 head, many of which were calves. Lorimet also had cattle mixed with Angorot's herd. By this time Lorimet's cattle had increased to about 30 in number, and from these he lost 15 during the Pokot raid. I heard later that three separate groups of Pokot attacked the Ngisonyoka cattle camps and took over 5,000 head of stock during this raid.

This entire period represented a time of stabilization and recovery for both Angorot's and Lorimet's livestock—that is, until the Pokot attack. Even though this time period was relatively dry, it is clear that all livestock species fared well, especially for Angorot. His camel holdings more than doubled, while his cattle and small stock holdings tripled. Lorimet's livestock holdings did not fare as well, mostly because he had to sell animals to buy food. He also occasionally slaughtered an animal to supplement the food that was purchased and the food that was given to his family by Angorot. The Pokot raid was a serious setback for these cattle herds. They had, for the most part, recovered from the losses sustained during Lochuu, but both herd owners emerged from this period with no more cattle than they had when the drought of 1980–81 ended.

WET SEASON 1985 TO WET SEASON 1987: FROM RECOVERY TO GROWTH

Following the Pokot raid on the cattle camps, Angorot and Lorimet made a series of rapid moves north along with the remaining cattle.

While the awi was near the town of Katilia, the cattle again turned south, this time moving toward the Suguta valley. The wet season of 1985 began in April and extended through mid-August. Although the wet season was considered good in terms of rainfall and the amount of primary productivity that resulted from the precipitation, many of the young small stock died of a disease called *Loutokoyen*, which gave that season and year its name.

During this time, Angorot and Lorimet remained in the eastern lava plains north of the town of Katilia, moving only a few times. This area contains very patchy forage resources, but during periods of good rainfall, annual grasses flourish.

Although at this time Lorimet was just beginning to get back on his feet, he chose to marry another wife. He did not have any livestock to transfer in bridewealth for Etir, his new bride, and arranged with her father to give animals at a later time. I mentioned in chapter 6 how surprised I was by this, and equally surprised by the opinion of the older men that this was one of Lorimet's best decisions. I will return to the relationship of herd growth and family formation in chapter 9.

In addition to a new wife, Lorimet also welcomed the return of his brother Lengess. Lengess had left Turkana District in the mid-1970s following a Pokot attack where he was shot in the leg. He worked as a laborer and herdsman on a farm outside the old colonial town of Kitale, in the Kenyan highlands. Lengess said that he worked along with other Turkana on a number of farms, but that he never learned kiSwahili and was treated "like a hyena" by many of the highland people. Lorimet sent word to Lengess that he now had enough animals to support him and that he needed the labor. Lengess was happy to return to Turkana after nearly ten years and said that he hoped that he would never again have to work for others outside of Turkana.

Only a trace of rain fell between August 1985 and March 1986, making this period the driest eight months of the decade. The Ngisonyoka had two names for this season/year. The first was *Epeipei alonyang*, referring to the yellow maize that was distributed by the government as famine relief food. This was the first time during the study period that this had happened, and it did not occur again until the early 1990s. The second name, *ekaru alolingakori*, referred to the dog flies that were unusually numerous; these flies were especially troublesome for the camels, biting and sucking their blood.

Although extremely dry, the Ngisonyoka did not consider this period a very bad dry season, due to the fact that enough forage grew following the rains in May and July to last well into the dry season. However, what really distinguished this period of time was the "operation" by the Kenyan army to disarm the Pokot. Newspaper reports suggested that some politicians from West Pokot District were being perceived by the central government as acting too independently and that they needed to be reminded of who was in charge. Regardless of the exact reason for the disarming campaign, the fact was that the Pokot were temporarily not such a serious threat.[3]

During the months of August through December, Angorot and Lorimet remained in the eastern lava plains, moving only three times. In January 1986, Angorot and Lorimet moved back into the Toma because there was lots of dry grass there, and they remained in the same location for two months. The fact that Angorot and Lorimet could remain in one place during the height of the dry season attests to the fact that the grazing resources were sufficient, although very dry. The nonmilking goats, however, remained behind in the eastern lava plains so that they could still have access to the acacia pods ripening along the dry riverbanks. The cattle remained grazing in the dissected hills in the far south of Ngisonyoka territory.

The wet season of 1986 lasted from March–April until September or October. Rain fell in each of these months, and in fact, the rains that began in March ushered in a period of above average rainfall lasting for the next three and a half years. Just as a period of little rain may not be as bad as the precipitation record indicates, a wet season with a lot of rainfall may not be as good as expected. This wet season was such a period. During this wet season an unusually large number of grasshoppers hatched, and as they grew they consumed much of the new plant growth. The season/year is called *ekaru edete,* the year of the grasshopper.

Angorot remained in the Toma until August and then began moving south into the hills of Nadikam in September. Once in Nadikam, he remained in the same awi until March 1987. This is the longest time in the entire study period that Angorot stayed at any one location. Angorot said that there was lots of dry season forage for his livestock in the immediate vicinity, and they were close enough to the Kerio River so that watering animals was not problematic. The nonmilking small stock had moved, and cattle remained separated from the main awi.

In the beginning of March, Nakwawi was married. Angorot transferred the first part of the bridewealth payment of ten camels to the father of Nakwawi's new bride. Once the rains began, all the small stock returned to the awi. This was also the time when Lorimet separated from Angorot. Lorimet felt that he now had enough animals to support himself and his family, and Lengess and his children were now back at the awi. Both Angorot and Lorimet insisted that the decision to separate in no way suggested ill will between the two herd owners.

During these years, the livestock herds for both Angorot and Lorimet grew significantly. Angorot's camels increased by 10, but Lorimet's camel herd decreased slightly due to the transfer of animals in bridewealth. Angorot's cattle herd more than doubled in size, while Lorimet's herd grew from basically nothing to 23 head. Angorot's small stock remained fairly stable, numbering around 1,250 animals, while Lorimet's flock increased slightly, from 70 to 84 goats and sheep.

WET SEASON 1987 TO DRY SEASON 1990:
A TIME FOR MARRIAGES

From here on Angorot and Lorimet resumed independent migratory patterns. The rains lasted from March through July 1987, and the Ngisonyoka considered this season a good wet season. The southern Ngisonyoka named this season/year *ekaru Agitomee,* the year of the elephants. This refers to the large number of elephants that migrated north that year from the Naroo into Nadikam.

Both Angorot and Lorimet remained in the Toma making only a few moves. The rains had stopped by August, and it remained dry until the following January. This was a relatively short dry season and was called by the Ngisonyoka *ekaru edira,* meaning "straight." This referred to the long straight swaths cut through the bush by oil company exploration teams. It was suspected that oil might be found in Turkana District, and for about six months a number of camps were set up from which oil company workers cut roads and set off small seismic charges. Oil was found but deemed too expensive to extract. The Ngisonyoka never did understand what was going on, but the oil companies left a number of very straight bush-free "roads" throughout southern Turkana.

Lorimet liked the area around the slow-flowing Kering wells, on the eastern side of the Toma, and because he had relatively small livestock holdings he could move to an area where there was limited but sufficient water for a small number of animals. He stayed in this vicinity for much of the dry season, but by February he moved south into Nadikam.

Like Lorimet, Angorot also moved south into Nadikam, moving short distances to areas where he had seen a small amount of rain falling. It is hard to pinpoint exactly when the dry season of 1987–88 ended and the wet season of 1988 began. It had rained in January, February, and March, but by April the rains began to get heavier and continued until December. The months of June, July, and October saw the heaviest rains. This wet season was given two names: *nadokichok*, referring to drips (meaning that the rain kept falling in small amounts throughout the period), and *ngakuutasinei*, meaning marriage (a large number of marriages took place during this season). This is a clear indication that the season was very good, as Turkana generally finalize marriages during seasons when the livestock are in good health and are giving a lot of milk.

For most of the next two years both Angorot and Lorimet followed a similar migration pattern. Rainfall was about average, and there were no raids by either the Pokot or Ngoroko. These years were not given names, which is an indication that life went on as normal, with no extreme events either ecologically or politically impacting people's lives. During these years both Angorot and Lorimet spent the wet season in the Toma and part of the dry season in Nadikam. Angorot kept his nonmilking small stock separate, but within a couple of kilometers of the awi, and they moved in concert with one another. Angorot's cattle remained far from the awi, in the southern range of Ngisonyoka territory.

This fairly regular migration pattern began to change in 1990, as it was drier than it had been in a number of years. Only a little rain fell during March and April, and by July the vegetation was completely desiccated. Angorot moved all the way south, close to the boundary with the Pokot. Lorimet also moved south, but just to the southern part of the Nadikam.

Some rain, but not much, fell during the remaining months of 1990, and although Angorot could not know it at the time, the dry season of 1990 was the beginning of a drought that proved to be worse than Lochuu. By September Angorot had crossed the Kerio River, a very unusual move for this early in the year, but an indication that the forage resources were in extremely short supply.

Up until this time the livestock for both herd owners continued to increase. Angorot's cattle herd nearly tripled in size, and his camel herd almost doubled. As the size of Angorot's family grew (see chap. 9), he sold goats and sheep frequently to buy maize and slaughtered many for food. He also transferred about 100 small stock in bridewealth payments. The size of his flocks of sheep and goats remained about the same.

Lorimet's livestock fared very well during this three-year period. His camels increased from 16 to 22, his cattle from 31 to 45, and his small stock from 130 to 300. Unfortunately for Angorot, Lorimet, and most other Ngisonyoka the good times of the late 1980s were about to change. Along with the drought came a renewal and transformation of Pokot raiding. As mentioned in chapter 5, the Pokot began to be supplied with arms by businessmen and some politicians. Livestock captured in raids were not brought back to the raider's homestead, but were transported south, to be sold in the markets of Nakuru, Nairobi, and Mombassa. Although I was not in Turkana District for the next six years, I was able to visit with both Angorot and Lorimet again in 1996.

Conclusion

Before turning to the analytical chapters I want to reiterate that I have been trying to provide the reader with a context for understanding the mobility and decision-making data. Condensing a ten-year period into a single chapter necessitates that a lot of detail is left out, but hopefully enough has been provided so that the reader appreciates how complicated decisions are, regardless of the answer given in response to a question about why a move was made. Fear of enemy raids, pressure from Ngoroko, the ever-present possibility that a dry season may lead to prolonged drought, the location of friends and relatives, the patchiness of the forage resources, and the predictions of emerons are all weighed and evaluated in each decision.

I will now turn to the analysis of the data relating to mobility patterns. I will be looking for variability as well as patterning and hope that by combining the ethnographic detail provided here with the analysis provided in the next chapter a more robust understanding of pastoral mobility will be achieved than has been possible in the past.

PART 3

CHAPTER 8

Data Analysis

Having given a narrative description of mobility and decision making, I next shift to an examination of what the data can tell us. One goal is to explore the extent to which Turkana mobility conforms to what is expected by viewing arid lands as nonequilibrium ecosystems. Ultimately, this analysis will allow us to examine the paradigm's usefulness in explaining the behavior of Turkana pastoralists.

If we are to understand arid land ecosystems as disequilibrial ecosystems then we would expect a high degree of variability and opportunistic mobility patterns. Opportunistic movements, those that take advantage of temporary patches of resources and avoid hazards, should be reflected in measures of variability on an annual and seasonal basis as well as among herd owners. One would also expect patterning in how the environment is used under particular climatic conditions. For example, as conditions become drier, people and livestock move into areas of progressively higher primary productivity. The precise location of this area may change every year, but the characteristics will be similar. In other words, it is the combination of variability and patterning that is the key. This is the challenge posed in the data analysis at the level of the individual.

Other issues will also be explored. Do herd owners move more frequently with increasing desiccation? More broadly, to what extent are individual movements a response to environmental conditions at all? How do security and markets influence pastoral mobility? These are some of the questions raised in this chapter.

Before turning to the data, I want to briefly reiterate my perception of the decision-making process as complex and at times idiosyncratic. Angorot is more daring than Lorimet and willing to take more risks. Lorimet has fewer livestock and can occupy places that have enough forage for his animals but that would not be suitable for a herd owner with larger livestock holdings. Lopericho chose to rely primarily on his camels for his family's subsistence, while his cattle remained separated from the awi. Atot has had to rely on the labor resources of his elder brother, and their personal relationship has at times been difficult. Much of this has been better captured in the preceding narrative chapters. Other patterns and behaviors will only be revealed by the analysis that follows.

SUMMARY STATISTICS AND ANNUAL VARIABILITY

The presentation of summary statistics allows for comparisons of herd owners. It is also necessary if comparisons are to be made among the Turkana and other pastoral peoples.

Angorot

The data set for Angorot spans ten and a half years, and during this time his awi moved a total of 1,301 kilometers in 119 separate moves. Angorot averaged 11.8 moves per year with an average distance of 10.93 kilometers per move. During the 126-month span, he remained in a single location for an average of 31.9 days before changing location.[1]

Angorot split his herds to a greater degree than the other herd owners. During the study period, Angorot kept his cattle separate 95 months out of the total 126 months. They moved 57 times during this period with an average distance per move of 16.25 kilometers. The cattle camp remained in a single location an average of 50 days before moving to a new location.[2] For quite a bit of this time the cattle were fairly close to the awi (about .5 kilometer) but in a separate camp.

Angorot kept his nonmilking small stock separate for approximately 31.3 months during the study period. During this time, they traveled 514 kilometers in 30 individual moves, with each move averaging 17.1 kilometers. The nonmilking small stock remained, on the average, 31 days in each location. Angorot also split off a very weak flock of nonmilking small stock from the satellite flock for about three months. This flock traveled 102 kilometers in three moves, averaging 34 kilometers per move and remaining in a single location approximately 28 days before moving.

Angorot split off his nonmilking camels when he felt it necessary. They were separate for a total of 21 months during the study period, but this was restricted to the period 1980 through 1984. The nonmilking camels traveled 253 kilometers in 13 different moves, averaging 18 kilometers per move. The average duration for the nonmilking camels to remain in one location was 48 days. There were a number of months where the camels were in a separate camp, but very close to the awi; these were not included in the calculations. A summary of all these data can be found in table 8.1.

One of the more hotly debated issues in the pastoral literature has been explaining the causes for pastoral mobility (chap. 2). In order to better understand the decision-making process, each herd owner was

TABLE 8.1.   Summary Statistics for Movement: Angorot (10 years) and Lorimet (85 months)

	Total Distance (km)	Mean Distance (km)	Number of Moves	Time Apart from Awi (days)	Average Duration of Stay (days)
Angorot					
Awi	1,301	10.93	119	—	24
Cattle	926.5	16.25	57	2,850	50
Nonmilking small stock	514	17.13	30	939	31
Weak nonmilking small stock	102	34	3	85	28
Nonmilking camels	253	18	13	630	48
Lorimet					
Awi	381	7.8	49	—	52
Cattle	397	39.7	10	855	85
Nonmilking small stock	—	—	—	—	—
Nonmilking camels	161	13.4	12	288	24

Source: Author's original data.

asked the most important reasons for moving to an area and for moving away from an area. During the early 1980s, I conducted this work
in conjunction with Rada Dyson-Hudson, and from our own experience we knew that decisions relating to moving away from an area
were often different from those relating to moving to a particular area.
We interviewed herd owners about these decisions until 1982, and I
continued this work throughout the rest of the study period.

Before presenting the data, I want to add a cautionary note.
Although the following data summarize the reasons given by Angorot
for his moves, issues of security and forage often are conflated. For
example, in December 1981, Angorot stated that the reason he moved
away from Komokuny was fear of Pokot, and the reason he moved to
Kakulit was because there was forage available for his livestock there.
Clearly the reason for the move to Kakulit was a combination of security (putting his family and herds farther from the Pokot) and environmental conditions at Kakulit. There were, no doubt, other locations
where forage was available, but these were perceived as more dangerous. However, the data recorded the reason for the move to Kakulit as
"forage available." I have also collapsed some responses into a category described as "forage." Most of the responses in this category were
only "forage" or "grass," but a few were more specific, such as "leaves
on shrubs for baby goats." The "miscellaneous environment" category
included reasons given for environmental concerns, but occurred only
once in the data set. Examples of this were "too much mud" and "too
many stones which are hurting the feet of the cattle." The data are summarized and presented in table 8.2.

I will discuss the implications of this data analysis in more detail
later when I compare the four herd owners, but I do want to point out
the overwhelming importance of the forage conditions for Angorot in
deciding where to move and the fact that security was also a major
factor in decision making, especially with respect to moving away
from particular areas. It was surprising that proximity to water was
rarely listed, and that access to goods and services never came up as
one of the most important factors in the decision-making process.

Annual Variability
The aggregate data summarized in the preceding sections allow us to
compare Angorot's mobility patterns with other herd owners' and to
those reported in the wider pastoral literature. However, the extent to

which these data vary, both annually and seasonally, may be equally important in addressing some of the important ecological questions we are asking. I next examine how key variables related to Angorot's mobility varied from one year to the next. These data are summarized in table 8.3.

TABLE 8.2. Reasons for Movement over a 10-Year Period for Angorot (% of total in parentheses)

Reasons for Movement	Movement To	Movement Away	Total Movement
Forage	81 (72)	62 (60)	143 (66)
Water	3 (3)	2 (2)	5 (2)
Security	8 (7)	23 (22)	31 (14)
Social/adakar	5 (4)	1 (1)	6 (3)
Moving to home area	4 (4)	6 (6)	10 (5)
Rain	9 (8)	0	9 (4)
Miscellaneous environment	3 (3)	10 (10)	13 (6)
Total	113	104	217

Source: Author's original data.

TABLE 8.3. Summary Statistics for Angorot by Year, 1980–90

	Distance (km)	Number of Moves	Average Distance per Move (km)	Duration of Stay (weeks)	Number of Weeks Cattle Separated	Number of Weeks Nonmilking Small Stock Separated	Number of Weeks Nonmilking Camels Separated
1980	168	14	12	3.4	30	11	12
1981	110	14	7.6	3.4	17	8.5	15
1982	191	15	12.7	3.2	0	0	0
1983	174	14	12.4	3.4	20	17	40[a]
1984	83	10	8.3	4.8	52	14	20
1985	214	16	13.4	3	52	35	0
1986	97	8	12.1	6	52	28	0
1987	112	9	12.4	5.3	52	0	0
1988	31	6	5.2	8	52	12	0
1989	82	8	10.3	6	52	0	0
1990[b]	56	5	11.2	4.4	22	0	0

Source: Author's original data.
[a]The nonmilking camels were separate but were close to the awi and moving with the awi.
[b]Data for 22 weeks only.

It is fairly clear that substantial variability is reflected for all the variables included in the table. The distance that the awi moved varied from a high of 214 kilometers in a single year to a low of 31 kilometers. The number of moves per year varied from a high of sixteen to a low of five. However, this variability appears to be couched within a larger multiyear pattern. Looking at the data in table 8.3, it is clear that the pattern of Angorot's moves can be grouped into two periods: 1980–85 and 1986–90. From 1980 through 1985 Angorot moved more frequently, for longer distances, and remained in each location less time than he did in the later period. He also kept his non-milking livestock separated for much more time each year. He kept his cattle separated from the awi more in the latter period, but here he often instructed Erionga to build the cattle camp only a short distance away from the awi and to move when the awi moved.

The most logical explanation for the variability between the two periods relates to rainfall. Figure 8.1 depicts the annual rainfall for the Ngisonyoka area from 1980 to 1990. The 1980–85 period was obviously far drier than the 1986–90 period. In fact, during the 1980–85 period, two severe droughts hit all of Turkana District.

This analysis would suggest that Angorot's mobility pattern is highly correlated with the precipitation pattern. On the surface this makes good sense. The drier it is, the more difficult it is to locate good forage for livestock, and thus we would expect to see more frequent moves as small patches of forage are exploited. However, there is another explanation which cannot be discounted, and that is security. The 1980–85 period was also characterized by frequent and intense raids, especially by the Pokot raiding across the border to the south. From previous chapters, the reader should be aware of how raiding has impacted the four herd owners and how it disrupted migratory cycles.

It may be that periods of more frequent raiding are associated with dry periods, as Bollig (1990) argues, for the Pokot. Researchers who advocate understanding arid lands as nonequilibrial ecosystems have suggested that raiding and violence are expected behavioral outcomes of living in this kind of environment. I will argue later that the long-term data for climate and raiding in Turkana District during the twentieth century do not support this position. It is often difficult to disentangle incidences of raiding from climatic influences. It seems logical that as livestock die from drought or disease, neighboring pas-

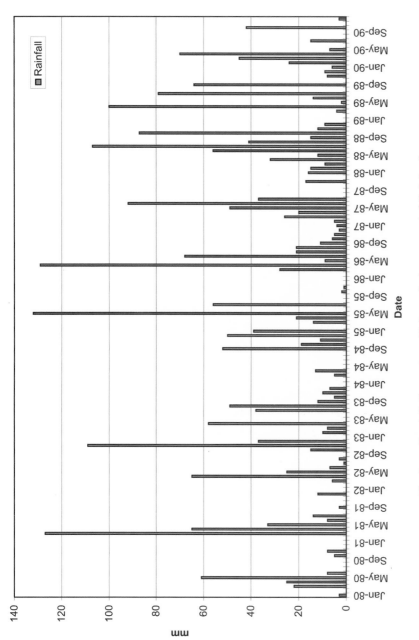

Fig. 8.1.   Precipitation pattern, Ngisonyoka (ten-year rainfall, Lokori)

toralists will try to recoup their losses through raiding. But it should be remembered that the Ngisonyoka refer to 1979 as the year of the GSU (*ekaru GSU*). At that time the government forces, the GSU, undertook a large "operation" designed to disarm and punish the Turkana for raiding. The same treatment was not inflicted on the Pokot, which left the Turkana vulnerable to Pokot raids. In the mid-1980s the Pokot were disarmed, and the amount of raiding decreased dramatically in southern Turkana. Trying to separate the influence of climate and that of larger political events may be impossible, at least in this context. I do not think that climatic conditions necessitate raiding, but for the Ngisonyoka the two appear to be interwoven, at least during the 1980s.

An examination of how Angorot responded to similar climatic conditions in different years reveals little consistency or patterning. The data for 1988 support the thesis that mobility patterns are highly related to rainfall. The year 1988 was the wettest of the decade, and Angorot moved much less than at any other time during the study period. Almost as much rain fell in 1986 and 1989, and again Angorot moved fewer times and for less distance than his average. I also know from accompanying him at this time that he felt that the environmental and security conditions were such that there was no need to move. However, 1985 was also a reasonably wet year, and Angorot moved sixteen times, totaling 214 kilometers, compared to eight times, totaling 82 kilometers, in 1989.

The same variation in response to similar ecological conditions also pertains to drought. The year 1980 had a very bad drought, and Angorot responded by moving frequently. Although 1984 was also a drought year, Angorot moved far less (ten times for 83 kilometers versus fourteen times for 168 kilometers). It may be that analysis of variability on an annual basis is masking patterns that will emerge when the data are examined seasonally. This possibility will be explored later in the chapter.

Lorimet

The data for Lorimet break down into two time periods. The first is when Lorimet was separate from Angorot, and all decisions relating to herd management were his own. The second refers to the period of time when Angorot and Lorimet were traveling together. This, however, was not an equal partnership. In March 1982, because Lorimet had lost nearly all of his livestock and could not survive on his own, he joined with Angorot. Angorot and Lorimet are brothers-in-law, and Angorot allowed Lorimet to share food and labor responsibilities. This allowed Lorimet and his family to recover from the drought and raids that had devastated his livestock. Lorimet continued to live with Angorot until September 1986; at that time Lorimet felt that he had recovered sufficiently to again live on his own. During the time that they were together, the main decision maker was Angorot. He would discuss movement options with Lorimet, but the decisions regarding when and where to move were primarily Angorot's. Lorimet always had the opportunity to split from Angorot and join with someone else or to go it alone. These events add complications to the analysis of Lorimet's data. Therefore, I have included the total data set, the time that Lorimet was separate, and the time when Lorimet and Angorot were together.

During the ten and a half years for which I collected data from Lorimet, the awi moved a total of 1,089 kilometers in 107 individual moves. The mean distance per move was 10 kilometers, and he remained at each location an average of 35 days before moving. For the 85 months when Lorimet was on his own, he traveled 381 kilometers in 49 separate moves. The mean distance per move was 8 kilometers, and he remained at each location an average of 52 days before moving to a new location.

Lorimet did not separate his livestock to the extent that Angorot did. During the 85 months that Lorimet was by himself, his cattle were separated a total of 28.5 months and his nonmilking camels for 9.5 months. During this period the cattle moved ten times for an average distance of 39.7 kilometers per move and remained in a single location for 85 days before moving. Lorimet's nonmilking camels moved a total of twelve times for 161 kilometers during the time that they were separate. They remained an average of 24 days in each location before moving. Lorimet never separated his small stock from the awi, but he lost almost all of his goats and sheep in a Pokot raid in December 1981. The aggregate data for Lorimet when he was on his own are summarized in table 8.1. I have not included the data, in this instance, for the period when he was with Angorot because for most of this time Lorimet had only a few livestock and was not making decisions on his own.

A labor shortage was one reason that Lorimet did not separate his nonmilking livestock to the extent that Angorot did. Lorimet had two brothers, Lengess and Achuka. Lengess migrated in the mid-1970s to the Kenya highlands to work on a farm following a Pokot raid in which he was shot in the leg. Achuka was a big, strong young man when I began working with Lorimet. He should have been ideal for herding either cattle or nonmilking camels. Unfortunately for Lorimet, Achuka was very irresponsible. When he was in charge of the nonmilking camels he would often leave them for days or weeks at a time, expecting some friends or relatives to take care of them. He would go to town, visit girlfriends, and drink local beer and a type of strong distilled liquor called *chunga*. In 1982 he began to accompany bandits (Ngoroko) while they stole livestock from Ngisonyoka. Lorimet often had to repay these animals to the families who had lost livestock, prompting Lorimet to say to me, "The biggest enemy I have is my brother, he is destroying all the livestock." In 1983 Lorimet asked the old men to help him with his brother. Following an all-day meeting and much debate, a decision was reached that required Achuka to leave Turkana District. If he refused he could be killed. He left and moved to the Kenya highlands near Eldoret where he remained until the early 1990s.

Table 8.4 summarizes Lorimet's reasons for movement during the time that he was on his own and when traveling with Angorot. Lorimet's reasons for movement demonstrate both similarities and some differences with those of Angorot. Environmental reasons accounted for over 80 percent of the moves, but unlike Angorot, availability of water figured prominently in Lorimet's decision making. Security appears to be less important to Lorimet in determining where to move than to Angorot, but these data are deceiving. Lorimet often followed far behind the leading edge of Ngisonyoka when they were moving south (thus closer to the Pokot). It is frequently said that these families form a shield for other Ngisonyoka following behind. Angorot was often part of this shield, while Lorimet was never among the families leading the movement to the south. Thus, security was less of an immediate concern when responding to questions concerning the determinants of movement, but was part of his overall mobility strategy.

Annual Variability

Data summarizing the key variables in Lorimet's annual mobility pattern are presented in tables 8.5 and 8.6. The patterning seen in the data for Angorot also appears in the data for Lorimet and is even clearer when the time spent with Angorot is added to the data set. Lorimet, however, responded quite differently than Angorot to both very wet and very dry years. Whereas Angorot moved frequently and traveled far during the 1980 drought, Lorimet moved nine times, but traveled only 71 kilometers. On the other hand, during 1988, the wettest year, Lorimet traveled farther and more frequently than at any other time, with the exception of 1981, another relatively wet year. The data also show that Lorimet did not move very far throughout the 1986–90 period. His small livestock holdings allowed him to stay in areas where herd owners with larger holdings could not. He and his cattle herd also remained far away from the Pokot. This strategy was successful in that he suffered no losses during this time from raiding, while others, including Angorot, did. Lorimet continued this strategy in the 1990s, and even though this was a period of great insecurity, Lorimet had avoided raids, and his livestock holdings grew while many others declined.

TABLE 8.4. Reasons for Movement over a 10-year period for Lorimet, Both Alone and with Angorot (% of total in parentheses)

Reasons	Lorimet To	Lorimet and Angorot To	Lorimet Away	Lorimet and Angorot Away	Lorimet Total	Lorimet and Angorot Total
Forage	25 (53)	117 (75)	19 (50)	49 (48)	44 (52)	166 (64)
Water	7 (15)	8 (5)	4 (11)	4 (4)	11 (13)	12 (5)
Security	2 (4)	7 (4)	7 (18)	26 (25)	9 (11)	33 (13)
Social/adakar	5 (11)	10 (6)	4 (11)	4 (4)	9 (11)	14 (5)
Home area	2 (4)	3 (2)	1 (3)	2 (2)	3 (4)	5 (2)
Rain	6 (13)	8 (5)	0	8 (8)	6 (7)	16 (6)
Miscellaneous environment	0	3 (2)	3 (8)	10 (10)	3 (4)	13 (5)
Total	47	156	38	103	85	259

Source: Author's original data.

Atot

During the 29.5 months for which I have data, Atot's awi moved a total of 335 kilometers in 32 separate moves for an average of 10.5 kilometers per move, and remained an average of 3.7 weeks in each location before moving. For the first year and a half, Atot's cattle were joined with those of his elder brother Loceyto. These cattle were divided into two herds under the management of two of Loceyto's sons, Epakan and Naperit. At the end of the first year, following significant losses due to drought and disease, the two herds joined together, and Atot brought his cattle back to the awi in May 1981. Thus, his cattle were separated for a total of 16.5 months. Epakan's herd traveled a total of 207 kilometers in 13 separate moves, for an average of 16 kilometers per move; this herd remained in a single location for an average of 5.1 weeks before moving. Naperit's herd was separated for a total of 11 months and during this time traveled at total of 266 kilometers in ten moves, for an average of 26.6 kilometers per move; this herd remained in a single location for an average of 4.4 weeks before moving. These data are summarized in table 8.7.

TABLE 8.5.   Summary Statistics for Lorimet by Year, 1980–90

	Distance (km)	Number of Moves	Average Distance per Move (km)	Duration of Stay (weeks)	Number of Weeks Cattle Separated	Number of Weeks Nonmilking Small Stock Separated	Number of Weeks Nonmilking Camels Separated
1980	71	9	7.8	5.7	52	0	0
1981	198	17	11.6	3.1	22	0	0
1982[a]	18	3	6	4.3	0	0	0
1983[b]	—	—	—	—	—	—	—
1984	—	—	—	—	—	—	—
1985	—	—	—	—	—	—	—
1986[c]	10	2	5	8	16	0	0
1987	21	5	4.2	10.4	28	0	0
1988	72	7	10.2	7.4	0	0	0
1989	33	5	6.6	10.4	0	0	0
1990[d]	62	5	12.4	4.8	0	0	0

Source: Author's original data.

[a]During 1982 Lorimet was separate from Angorot for just over 12 weeks.
[b]During this time Lorimet was traveling together with Angorot.
[c]During 1986 Lorimet was separate from Angorot for 16 weeks.
[d]The data for 1990 includes 22 weeks.

TABLE 8.6.   Summary Statistics for Lorimet (including the time with Angorot) by Year

	Distance (km)	Number of Moves	Average Distance per Move (km)	Duration of Stay (weeks)	Number of Weeks Cattle Separated	Number of Weeks Nonmilking Small Stock Separated	Number of Weeks Nonmilking Camels Separated
1980	71	9	7.8	5.7	52	0	0
1981	198	17	11.6	3.1	22	0	0
1982[a]	175	15	11.7	3.5	0	0	0
1983[b]	174	14	12.4	3.4	20	17	40
1984[b]	83	10	8.3	4.8	52	14	20
1985[b]	214	16	13.4	3	52	35	0
1986[c]	41	8	5.1	6.5	44	0	0
1987	21	5	4.2	10.4	28	0	0
1988	72	7	10.2	7.4	0	0	0
1989	33	5	6.6	10.4	0	0	0
1990[d]	62	5	12.4	4.8	0	0	0

Source: Author's original data.
[a]During 1982 Lorimet was separate from Angorot for just over 12 weeks.
[b]During this time Lorimet was traveling together with Angorot.
[c]During 1986 Lorimet was separate from Angorot for 16 weeks.
[d]The data for 1990 includes 22 weeks.

Atot separated his nonmilking small stock for a total of 38 weeks during the 29 months for which I have data. During this time the small stock traveled a total of 151 kilometers in ten moves, averaging 15.1 kilometers per move, and remained an average of 3.8 weeks in each location. Atot's responses about his reasons for movement are summarized in table 8.8. I am going to postpone commenting on Atot's (and Lopericho's) determinants of mobility and on the importance of annual variability until later in the chapter when all four herd owners will be compared with each other.

Lopericho

The data for Lopericho span 29 months, and during this time he traveled 406 kilometers in 35 separate moves, averaging 11.6 kilometers per move. Lopericho remained on the average 3.3 weeks in each location before moving. Lopericho had his cattle divided into two herds, each managed by a younger brother. The cattle herd managed by

TABLE 8.7.   Summary Statistics for Movement: Atot and Lopericho (29.5 months)

	Total Distance	Mean Distance	Number of Moves	Time Apart from Awi (days)	Average Duration of Stay (days)
Atot					
Awi	335	10.5	32	—	26
Cattle (Epakan)	207	16	13	464	36
Cattle (Naperit)	266	26.6	10	308	31
Nonmilking small stock	151	15.1	10	266	27
Nonmilking camels	—	—	—	—	—
Lopericho					
Awi	406	11.6	35	—	23
Cattle (Apuu)	686	22.1	31	870 (all)	26
Cattle (Komosiong)	190	38	5	476	95
Nonmilking small stock	?	?	?	504	?
Nonmilking camels	—	—	—	—	—

Source: Author's original data.

Kamosiong was separated a total of 68 weeks and during this time traveled 190 kilometers in five moves, averaging 38 kilometers per move, and remained 13.6 weeks in each location before moving. The other herd, managed by Apol, was separated throughout the study period. During this time, Apol's herd traveled 686 kilometers in 31 moves. Apol averaged 22.1 kilometers per move and remained for an average of 3.7 weeks in each location. These data are summarized in table 8.7.

Lopericho never separated his nonmilking camels, and once his small stock returned from their sixteen-month initial separation, they were only away from the awi for two additional months. I was not able to collect good data for this flock when they moved north as there was no contact between Lopericho and the small stock, and we could only reconstruct the movements of this flock in general terms when the flock returned.

Table 8.8 summarizes Lopericho's reasons for movement. As with

TABLE 8.8.   Reasons for Movement over a 29-Month Period for Atot and Lopericho (numbers in parentheses are percentages of the total for each herd owner)

Reasons for Movement	Atot			Lopericho		
	Movement To	Movement Away	Total Movement	Movement To	Movement Away	Total Movement
Forage	15 (63)	7 (29)	22 (46)	25 (81)	13 (45)	38 (63)
Water	1 (4)	1 (4)	2 (4)	0	0	0
Security	2 (8)	13 (54)	15 (31)	5 (16)	9 (31)	14 (23)
Social/adakar	4 (17)	3 (13)	7 (15)	0	0	0
Moving to home area	1 (4)	0	1 (2)	0	2 (7)	2 (3)
Rain	0	0	0	0	3 (10)	3 (5)
Miscellaneous	1 (4)	0	1 (2)	1(3)	2 (7)	3 (5)
Total	24	24	48	31	29	60

Source: Author's original data.

the data for Atot, I will wait to comment on them until later in the chapter.

Discussion

One of the key characteristics of nonequilibrial ecosystems is that migration patterns should demonstrate a high degree of variability on an annual basis. The analysis just presented illustrates this variability for these four herd owners. In order to explore this in a bit more detail, I summarized the data on mobility for the four herd owners on an annual basis and restricted the data to a twenty-nine-month period for which I have measures for each herd owner. These data are presented in table 8.9.

I have previously argued that the high degree of flexibility in livestock management practices is one of the key adaptive strategies of the Turkana (McCabe 1985, 1994). This variability is evident in the data summarized here. Not only is there marked variability among herd owners in any one year, but there is also substantial variability from one year to the next for individual herd owners. The data for Lorimet are instructive here. During 1980, a severe drought year, the other herd owners moved frequently and traveled almost twice as far

TABLE 8.9. Annual Variability of the Four Herd Owners, 1980–82

Herd Owner	Variables	1980	1981	1982
Angorot	total distance (km)	168	110	141
	number of moves	14	14	8
	mean distance (km)	12	7.6	17.6
	duration of stay (weeks)	3.4	3.4	2.8
	number of weeks cattle separate	30	17	0
	number of weeks nonmilking camels separate	12	15	0
	number of weeks nonmilking small stock separate	11	8.5	0
Lorimet	total distance (km)	71	198	119[a]
	number of moves	9	17	7
	mean distance (km)	7.8	11.6	17
	duration of stay	5.7	3.1	3.1
	number of weeks cattle separate	52	22	0
	number of weeks nonmilking camels separate	0	0	0
	number of weeks nonmilking small stock separate	16	22	0
Atot	total distance (km)	135	66	134
	number of moves	10	10	12
	mean distance (km)	13.5	6.6	11.2
	duration of stay	5.2	5.2	1.8
	number of weeks cattle separate	52	18	0
	number of weeks nonmilking camels separate	0	0	0
	number of weeks nonmilking small stock separate	20	18	0
Lopericho	total distance (km)	134	182	90
	number of moves	13	14	8
	mean distance (km)	10.3	13	11.3[e]
	duration of stay	4	3.7	2.8
	number of weeks cattle separate	52[b]	52[d]	52
	number of weeks nonmilking camels separate	0	0	0
	number of weeks nonmilking small stock separate	52[c]	20	0

Source: Author's original data.

[a]Part of this time Lorimet was traveling with Angorot.

[b]Lopericho had two herds of cattle, and both were separate for the entire year.

[c]The nonmilking small stock were separate for 2 months, and all small stock were separate for the remaining 10 months.

[d]One cattle herd rejoined the awi after 5 months, and the other remained separate for the entire year.

[e]Data for Lopericho from 1982 consist of 22 weeks.

as Lorimet. Lorimet clearly had a different response to drought than the others; this may have been possible because of his smaller livestock holdings or his view that the energy expenditures resulting from movement would not be balanced by access to better forage. Likewise, in 1981 Lorimet moved greater distances and more frequently than any of the other herd owners, again demonstrating a different management style and strategy.

Another way of examining variations among the herd owners is to look at the way they responded to questions about their reasons for moving (table 8.10). Although there are differences among the herd owners, the importance of access to good forage is apparent for each of the herd owners. Angorot and Lopericho did not move much to be close to other people, while Lorimet and Atot did. Security was a major issue for all the herd owners. What may be as interesting in this data set is what is not there. There is not a single response related to economic issues or wanting to be close to town. For decades, the extent to which access to markets, schools, and medical clinics influence pastoral mobility had been a subject of debate (McCabe 2000). The data here demonstrate that these factors were not of major importance to the Turkana people with whom I worked during the study period, and this confirms my own opinion based on living and traveling with these herd owners.

TABLE 8.10. Determinants of Movement Based on All Available Data for the Four Herd Owners Expressed as Percentage of Total Responses

Reasons for Movement	Angorot	Lorimet	Atot	Lopericho
Forage	66	52	46	63
Water	2	13	4	0
Security[a]	14	11	31	23
Social/adakar	3	11	15	0
Home area	5	4	2	3
Rain	4	7	0	5
Miscellaneous/environment	6	4	0	5
Miscellaneous	0	0	2	0

Source: Author's original data.

[a]Security issues were very important in the years 1980–82, and therefore the percentages for Angorot and Lorimet are understated here in comparison with Atot and Lopericho. In a previous publication where I used only the 1980–82 data I estimated that security made up 20% of total responses for Angorot and 24% of total responses for Lorimet (McCabe 1994).

What these data demonstrate in terms of variability, supported by the previous chapter's narratives, is that each herd owner has his own strategies for coping with environmental stress and variability. These are based on a unique perspective of risk, personal history, livestock holdings, and labor supply. Another way of examining variability and the extent to which individual herd owners follow a pattern is to break down the annual rainfall figures into particular seasons. After all, this is how the Turkana view the climatic pattern.

Seasonality and Movement

An analysis of mobility and decision-making data looking for patterns and variability on an annual basis may help support the argument that Ngisonyoka pastoralists use the land in a way that is consistent with an understanding of arid land ecosystems as nonequilibrium systems. But this is not the way that the Ngisonyoka, or other Turkana, view climate. For the Turkana, each season is named and has its own characteristics. The length of the season can vary from a few months to more than a year. I have information on how the Ngisonyoka viewed each season based on informal discussions with herd owners during the course of fieldwork and also from interviews conducted later. I have used this information but have taken my own approach to identifying seasons, taking into account the three-month moving average for precipitation, my own notes on climatic conditions, and what my friends and informants were telling me at the time. I have used a three-month moving average instead of the monthly precipitation here because it takes one to two months following a rain for the vegetation to grow and turn green. Likewise, it takes two or more months after the rain has stopped for the vegetation to dry out. The three-month moving average captures changes in the vegetation far better than the actual rainfall and thus is a much better indicator of forage conditions.

I have divided the period 1980 to 1990 into twenty different seasons (table 8.11). I have used these seasonal divisions in some of the analysis to follow, but I have also collapsed the data into seven and sometimes three climatic periods, depending on the analysis.

Table 8.12 summarizes data for the awis of the four herd owners for the period January 1980 through the first two weeks of July 1982, the period of time for which I have data for each herd owner. Duration of stay and number of moves per month are important because they indicate the management strategy for each herd owner. The seasons represented here are: Period 1, January 1980–March 1981 (1 in table 8.11); Period 2, July 1981–March 1982 (3 in table 8.11); Period 3, April 1982–July 1982 (12 in table 8.11); and Period 4, March 1981–June 1981 (16 in table 8.11).

TABLE 8.11.   Seasons in Ngisonyoka Arranged from Driest to Wettest, 1980–90

	Time Period	Characteristic
1	Jan. 1980–March 1981	extreme drought
2	Nov. 1983–Sept. 1984	drought
3	July 1981–March 1982	very bad dry season, but not drought
4	Aug. 1985–March 1986	very bad dry season, but not drought
5	Sept. 1987–June 1988	bad dry season
6	Sept. 1986–April 1986	bad dry season
7	July 1990–Dec. 1990	dry season
8	July 1982–Oct. 1982	dry season
9	Jan. 1989–March 1989	dry season
10	Dec. 1989–Feb. 1990	dry season
11	April 1983–Oct. 1983	poor wet season
12	April 1982–June 1982	poor wet season
13	Nov. 1982–Jan. 1983	wet period in normally dry time
14	March 1990–June 1990	poor wet season
15	May 1987–Aug. 1987	good wet season
16	March 1981–June 1981	good wet season
17	Oct. 1984–July 1985	good wet season
18	April 1986–Aug. 1986	good wet season
19	April 1989–Nov. 1989	very good wet season
20	June 1988–Dec. 1988	exceptionally good wet season

Source: Author's original data.

TABLE 8.12.   Duration of Stay in Weeks (DUR) and Moves per Month (MOV) for Each Herd Owner by Seasonality

	Period 1		Period 2		Period 3		Period 4	
	DUR	MOV	DUR	MOV	DUR	MOV	DUR	MOV
Angorot	3.9	1.03	4.0	1.0	2.0	2.0	2.33	1.71
Lorimet	6.2	0.65	3.0	1.33	2.0	2.0	2.8	1.43
Lopericho	3.9	1.03	2.57	1.56	4.67	1.17	7.0	0.57
Atot	5.64	0.71	5.14	0.78	1.75	2.29	2.8	1.43

Source: Author's original data.

Angorot, Lorimet, and Atot each remained longer in a single location during the drought period than at any other time. Lopericho did not follow the same pattern. He remained in a single location for approximately four weeks during Period 1, and his frequency of movement increased during Period 2 (this was also true of the other herd owners). However, during Period 3 and especially during Period 4 he decreased his movements dramatically. As we will see later, Lopericho appears to follow a movement strategy quite different from all the other herd owners'.

Another way of examining the data is to pose the question, Is there a significant relationship between seasonality and patterns of movement? I tested this by calculating the correlation coefficients for seasonality, number of moves per month, and duration of stay for the four herd owners from January 1980 through June 1982. With the exception of Lopericho, all of the correlation coefficients for the herd owners were significant. The data indicate that, in general, the herd owners remain in single locations for longer periods of time as precipitation decreases. I must admit that this finding challenges my impression of how Turkana mobility patterns were structured. I had always believed that during times of extreme dryness the herd owners that I knew, with the exception of Lorimet, moved more frequently and divided their herds into the smallest units possible. This analysis seems to suggest that this interpretation is wrong. However, another explanation may be possible, and I will consider that shortly.

Another, more complex, pattern emerges when the data for either duration of stay or moves per month are plotted against seasonality. Figure 8.2 illustrates this pattern for moves per month. Here we can see that three of the four herd owners increased their movements from Periods 1 through 3, with Period 1 being very dry and Period 3 being a poor wet season. Period 4, however, was quite wet and the frequency of mobility decreased. Why would the frequency of movement increase so much between periods 2 and 3, and then decrease in Period 4? I believe that the explanation is that during Period 3, forage was available in patches and the herd owners moved to exploit it. During Period 4, forage was more abundant and less patchy, and therefore there was less to be gained from frequent movement. What about Lopericho? He diverges dramatically from the others during Period 3, and I know that at that time Lopericho was traveling with only his camels. Camels can range farther during the day for forage, and patches can be exploited without moving the entire awi.

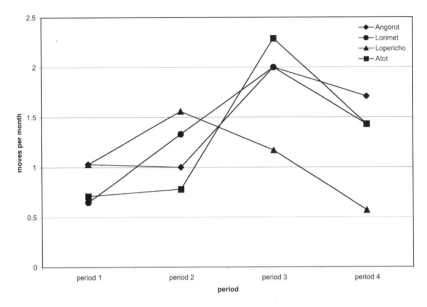

Fig. 8.2.   Moves per month for each herd owner by seasonality

Although it is somewhat unusual to have 29 months of consecutive data for nomadic peoples, it is still a small piece of the overall climatic pattern. An examination of the full ten years of data for Angorot and Lorimet reveals a somewhat different picture about their management patterns with respect to seasonality. I again ran correlations between seasonality and moves per month, but in this analysis no figures were significant, for either Angorot or Lorimet. The correlations for duration of stay were not significant for Angorot but were for Lorimet. Thus the influence of climatic conditions was less clear in the larger data set than in the smaller one. Nevertheless the interpretation one would arrive at based on correlations would be similar to that mentioned previously. Using the ten-year data set for Angorot and Lorimet, as well as that for seasonality, I plotted the number of moves per month against seasonality. This is illustrated in figure 8.3.

In this figure, Season 1 is composed of the two best seasons; Season 2 includes the five "normal" wet seasons; Season 3 includes the three poor wet seasons; Season 4 includes the four "normal" dry seasons; Season 5 includes the two bad dry seasons; Season 6 includes the two very bad dry seasons; and Season 7 is drought. During the best climatic conditions both Angorot and Lorimet moved infrequently. This makes

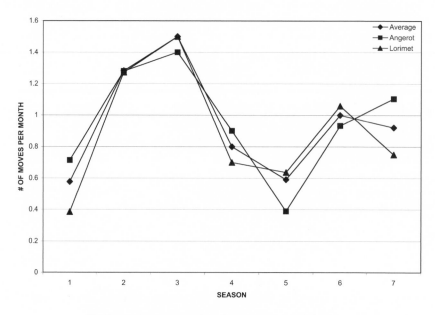

Fig. 8.3.   Average number of moves per month by season

sense, as there would be abundant forage and water nearly every-
where. I often asked herd owners about mobility during these times
and was frequently told that they felt no need to move, as everything
they needed was close by. During Seasons 2 and 3, mobility increased.
During these times forage would be less abundant, and Angorot and
Lorimet would be keeping all their livestock together. Keeping all the
livestock at the awi requires enough forage for each species in close
proximity to the awi. During poor wet seasons these conditions are
harder to find, and the resources are exhausted fairly quickly, thus
requiring more mobility. Seasons 4 and 5 are dry seasons, and we see
mobility declining fairly dramatically. Two things would be happening here. First, satellite herds would be in the process of being split off;
second the energy expenditure for each move may not be offset by
improved forage conditions. In Season 6 climatic conditions are very
bad, and mobility increases. At this point forage resources are limited
and very patchy. Livestock would probably be dying, and the herds
are most likely at maximum separation. Access to forage becomes critical, and mobility increases. During drought, nothing much can be

done. Angorot and Lorimet would be in the southern, more productive area of Ngisonyoka territory, but the vegetation would be totally dried out. Little would be gained by moving.

This analysis of frequency of movement suggests a complexity and patterning of pastoral mobility not reported in the literature. The interplay of access to forage, on the one hand, and the energy expenditure resulting from movement, on the other, are trade-offs, and any decision to move must take this into account. Considering the variability seen in the other analyses, I was struck by the similarity among herd owners in this examination of mobility. It suggests an understanding of range ecology and livestock management on the part of herd owners not appreciated in previous research.

Seasonality and Location

Despite the variability observed on an annual basis, patterning seems to be characteristic of mobility with respect to frequency of moves and seasonality. It seems obvious that patterning would also be observable with respect to seasonality and location. It is certainly the case that individuals tend to move to the south each year as the dry season progresses. As previously mentioned, the south is wetter than the north, and this type of mobility pattern is consistent with the literature on wild animal migration that suggests migratory ungulates spend the wet season in areas of relatively low primary productivity and move to areas of high productivity during the dry season. The basic argument is that moving in this way allows more animals to be supported by the rangeland than either sedentarizing the ungulate population or using other migratory patterns (Behnke and Scoones 1993). Certainly one cannot equate or reduce the decision-making process of individual herd owners to that of wild migratory ungulates. However, if similar patterns are found, this may say something about the nature of Turkana land use, and possibly about migration itself.

In order to look for patterning beyond a north-to-south migration, I constructed a primary productivity map based on Coughenour's precipitation map. In arid areas primary productivity is highly correlated with rainfall; therefore a precipitation map could serve as a good proxy measure for primary productivity. I broke Ngisonyoka territory into five categories with 1 being the area of lowest primary productivity and 5 being the highest. I superimposed all locations for each herd owner on the primary productivity map and categorized the productivity for each location. I then went back to my field notes and, combined with rainfall data, designed a vegetation index from 1 to 5, with 1 being very good conditions and 5 being exceedingly dry conditions. I correlated the vegetation index with the primary productivity category for each location for all herd owners.

A significant relationship at the .05 level was found between the vegetation index and the level of primary productivity for all herd owners. The correlations are not as strong for Lopericho and Lorimet (not significant at the .03 level). The data certainly suggest that as the vegetation dries out the herd owners move into areas of higher primary productivity. This is not surprising; except for the fact that it occurs on a regular basis and that the relationship is statistically significant for all herd owners. The correlations for Lopericho are significant, but the correlation coefficients were negative. Lopericho appeared to move into areas of lower productivity as vegetal condition worsened. One explanation for this may be that his herds consisted primarily of camels, and they were able to forage from the leaves of trees growing along dry riverbeds in areas where the overall vegetation index was low. I suspect that this is true, as the cattle and small stock were separate for most of the study period, and I know that Lopericho was often in areas where grazing animals would not have survived. Nevertheless, the data analysis for Lopericho was surprising and again suggests that mobility patterns and strategies can be highly idiosyncratic.

Conclusion

In the data analysis regarding mobility and decision making for the herd owners of the four families, both variability and patterning have been identified among the families regarding how they used the land, made decisions, and managed their livestock. However, this is only part of the picture. The relationship between the growth of the herds and that of the family may be as important in understanding the long-term strategy of individual herd owners as are mobility and the use of the environment's natural resources. This will be considered in chapter 10.

CHAPTER 9

Livestock Dynamics and the Formation and Growth of Families

Mobility among pastoral people is a strategy whereby individual herd owners attempt to maintain and improve the productivity and fertility of their herds as environmental and political conditions change. One goal is clearly to increase herd size—but to what end? Or is this the end in itself? This question has engaged specialists in African pastoralism for most of the twentieth century. The question of why African pastoralists keep such large herds was a topic introduced into the anthropological literature by Herskovits in 1926, and it continues to generate debate today. The early assumption, now discarded, suggested the keeping of large herds of relatively low productivity animals was an indication of the "irrationality" of African pastoralists. The "irrationality" explanation was countered by materialist arguments that centered on the need for large herds precisely because of their low productivity (Deschler 1965; Schneider 1957), or arguments stressing that pastoralists knew many animals would die during drought, and the larger the herd, the larger the core herd would be that formed the basis for recovery (Fratkin and Roth 1990). "Pastoralists store wealth on the hoof" is a quote from Legge (1989:83) used by Roth to capture this argument (Roth 1996:219). The underlying assumption here is that the herds were an end in themselves—they provided food and were a store of wealth. The fact that most African pastoralists were reluctant to slaughter or sell animals except when absolutely necessary reinforced the thinking that large herds were ends rather than means.

What was, and to a large extent still is, missing from these explanations is the goals of the pastoralists themselves. Are African pastoralists trying to maximize livestock numbers, and if so, are they doing this primarily as a means of storing wealth? In the following sections I am going to argue that herd size is indeed important for maintaining an adequate food supply, and that livestock populations do crash dramatically during drought, but the livestock herd is also the primary means by which individual pastoral people are able to initially form a family, and it is through the herd that family growth is possible. The decisions about if and when to transfer livestock for bridewealth are as critical as decisions about if and when to move—if not more so.

The data on livestock dynamics and family growth can also be used to assess the differential success of an aggressive mobility strategy (Angorot) versus a conservative strategy (Lorimet). The data on Lorimet documents how individuals recover from near-disaster, while that for Lopericho illustrates how a successful and wealthy family can become impoverished and disintegrate following the death of the herd owner.

IMPACT OF DROUGHT

The theory of arid lands as nonequilibrial ecosystems predicts that livestock numbers will go through cycles of crashes and recovery. Crashes are principally associated with environmental perturbations, and in tropical arid lands these will usually occur in the form of droughts. The literature is replete with accounts of livestock losses of 50 percent or more for African pastoralists (I have also published Turkana material on this topic). Here it may be useful to summarize this data and to frame it within the nonequilibrial ecosystem arguments.

Table 9.1 summarizes the aggregate data as well as data for each herd owner for livestock losses during the 1980–81 drought. Overall, herd owners experienced a loss of almost 60 percent of their livestock holdings, with the losses from cattle being the most significant. I have used tropical livestock units (TLUs) as a means by which the different sizes and productivity of each livestock species can be converted in a common unit. There are many ways of calculating TLUs but I have followed the formula used by Little and Leslie (1999). In this calculation a camel is valued at 1.25 TLU, cattle at 1.0 TLU, and goats and sheep at .125 TLU.

An examination of differential losses among individuals suggests that neither the size of the individual herd nor the management strategy seems to have played an important role in herd survival with the exception of cattle. Atot had, by far, the largest cattle herd, and a large percentage of his cattle died due to contagious bovine pleuropneumonia (McCabe 1987). Angorot's aggressive management strategy was not as successful as Lorimet's conservative strategy in withstanding the drought, but Lorimet's lack of mobility was made possible by his small herd size. Lopericho fared the best, which is not surprising as he based his management on the herding of camels, by far the most drought resistant of the livestock species.

Another drought occurred in 1984–85 but was less severe than the 1980–81 drought. It was of a shorter duration, and the southern part of Ngisonyoka territory did receive rain that year. I have data for both Angorot and Lorimet for this time period, and both suffered major losses, but due to raiding, not drought. Angorot's camels declined from 81 to 60 during this period, but he gave 23 camels in 1984 for the bridewealth of his brother Akopenyon's first wife. His small stock actually increased, but the Pokot raided all the cattle except calves.

TABLE 9.1.   Total Livestock Losses and Percent Losses by Herd Owner, 1980–81 Drought

Losses	Camels	Cattle	Small Stock	Tropical Livestock Units (TLUs)
1980	257	592	1,734[a]	1,130
1981	164	164	795	468
Angorot	34%	66%	59%	55%
Lorimet	70%	56%	59%	46%
Atot	42%	86%	36%	73%
Lopericho	30%	48%	49%	40%
Total % loss	36%	72%	54%	59%

Source: Author's original data.

[a]The 1980 numbers of Lopericho's small stock are an estimate based on interviews rather than actual counts.

The same pattern was characteristic of Lorimet's herds as well. His small stock increased, as did his camel herd, but he lost all his cattle in the same series of Pokot raids that affected Angorot.

Droughts of the magnitude reported on here have occurred in Turkana District on average once every ten years. If these data are representative of the overall livestock population, crashes of this magnitude could be expected at least once each decade. The theory of arid land ecosystems as nonequilibrium systems posits that because of periodic crashes in the herbivore populations, these populations will not be stable enough over time for close plant-herbivore linkages to develop and that environmental degradation will not be a problem because herbivores do not increase to the extent that a theoretical carrying capacity is reached. The data I present here based on a small sample lend support to this argument, and more convincing data are presented later (chap. 10, "Aggregate and Group Movement").

Herd Dynamics in the Longer Term

The impact of drought obviously plays a major role in determining the long-term dynamics of Turkana livestock populations. However, there has been very little multiyear research that has focused on how individual herds and flocks fare over periods as long as a decade. This type of research is necessary in order to contextualize drought within a larger framework. Table 9.2 presents summarized data for Angorot and Lorimet spanning ten and sixteen years respectively. The data are for the most part actual counts, but in a couple of instances, I had to extrapolate from the counts of herds based at the awi to separate satellite herds. Because the well-being of the family is ultimately a consequence of what happened to all the livestock, I have emphasized the changes in the combined herds as measured in tropical livestock units (TLUs). The changes in TLUs for the ten- and sixteen-year periods are illustrated in figure 9.1.

Despite two droughts and four raiding episodes, the overall herd numbers, at least as measured by TLUs, actually increased during the decade of the 1980s. Including the many animals that were also given in bridewealth during this time would have increased the numbers of animals surviving quite substantially. With the exception of the 1980–81 drought, the largest losses occurred because of raids. It

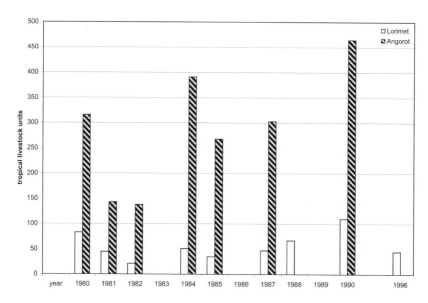

Fig. 9.1.   Angorot and Lorimet: TLUs, 1980–96

TABLE 9.2.   Changes in Herd Size for Angorot and Lorimet, 1980–96

	Angorot				Lorimet			
	Camels	Cattle	Small Stock	TLUs[a]	Camels	Cattle	Small Stock	TLUs[a]
1980	68	107	991	316	30	45	268	83
1981	45	36	405	143	9	20	113	45
1982	33	48	393	138	7	12	0	21
1983								
1984	81	165	1,000	391	13	30	35	51
1985	60	30	1,300	268	15	0	70	28
1986								
1987	70	65	1,200	303	11	23	84	47
1988					16	31	130	67
1989								
1990	126	157	1,200	465	22	45	300	110
1996					11	0	250	45

Source: Author's original data.

[a]Tropical livestock units

would be logical to assume that the losses due to raids were offset by inputs of livestock from raids, but this was not the case. Most of the Turkana raiding during the whole study period was conducted by individuals from the northern sections, even that waged against the Pokot and Samburu. There were a few instances in which some people that I knew joined in these raiding parties, but this was rare, and these individuals were not part of our study group. The Ngisonyoka were at a serious disadvantage because of the lack of rifles, and the Ngoroko who led these raids had little affection for the Ngisonyoka.

These figures attest to the resilience of Turkana livestock and the skill of the livestock managers. The number of livestock held fluctuated greatly, but the period between 1985 and 1990 were times of steadily increasing herds. Environmental conditions were good, and there was very little raiding following the disarming of the Pokot in 1986. However, the fortunes of both Angorot and Lorimet changed during the 1990s—dramatically for Angorot, and less dramatically for Lorimet. A sustained period of drought began in 1990 and was accompanied by new waves of even more violent and devastating raiding by the Pokot. I was able to count Lorimet's livestock in 1996, and these numbers are included here. I visited with Angorot and know what happened to his herds, so his overall herd size is an estimate (as I was not able to count his livestock).

I also want to caution the reader that these figures are only based on two of the four herd owners. I know that Atot was not able to build his herds back to the 1980 levels, and I will discuss the unfortunate events that led to the complete loss of Lopericho's herds. In addition, the figures demonstrate how an individual who has been made nearly destitute can recover if helped by others. Lorimet would not have remained in the pastoral sector if Angorot had not been willing to come to his aid. Angorot himself was only able to become a viable pastoralist because his mother's brother was willing to help him following the death of his father.

Losses Due to Raiding

Raiding is critically important in influencing decisions related to mobility (chap. 5). Despite their best efforts, all the herd owners in this study were severely impacted by Pokot raiding. Raiding in the years following his death devastated Lopericho's herds. Atot lost very few animals to raiding until 1992, when most of his cattle were lost in Pokot raids. Pokot raiding increased in frequency and ferocity throughout the first half of the 1990s, and Atot was again raided in 1995, losing his entire flock of nonmilking small stock. Angorot lost animals to raids throughout the entire study period. In 1980 he lost 350 goats and sheep, nearly the whole flock of milking small stock, and in 1985 he lost nearly all his milking cattle. In 1992 he lost 12 camels and 15 donkeys in raids, and in 1995 he lost 86 camels, most of these either pregnant or milking camels. This represents a tremendous loss of livestock capital for a single herd owner. Camels were selling at the time for between 9,000 and 10,000 Kenyan shillings each (at the time 1 U.S. dollar was worth 55 Kenyan shillings). Using 9,500 shillings as an average price this would equate to a loss of 817,000 Kenyan shillings.

Lorimet presents an interesting case in terms of managing his animals to reduce the risk of raiding. His strategy has always been to remain behind most of the families as they move closer to enemies, thus sacrificing increased access to forage for relative safety. This has
worked well except in two instances. The first occurred in 1981 when he moved his small stock to the Naro. This was an unusually bold move for Lorimet. Unfortunately it corresponded to the time of a series of large Pokot raids, and he lost nearly all of his goats and sheep, leaving him with his cattle and the few camels that remained in the Toma. In 1985 his cattle were lost in the same set of raids in which Angorot lost his cattle. However, from 1985 until 1996, Lorimet had not suffered any losses due to raiding.

Reports of destitution and starvation among the Turkana are frequently cited in the national and international press, and many people have had to depend on famine relief for survival. Drought and poor management are often blamed for this dire situation, but raiding, especially by the Pokot against sections of the southern Turkana, has been more devastating than any drought. Some authors have viewed raiding as an expected response among people living in nonequilibrium ecosystems. However, I do not agree with this position. Raiding came to a complete stop following the disarming of the Pokot in 1986 and did not resume until they were able to rearm themselves. The decades from 1920 to 1950 were drier than the decades of the 1980s and 1990s, but this was a time of peace among the pastoral peoples living in northwestern Kenya. Pressure to move into neighboring territories during drought certainly can lead to conflict, and raiding is a means by which individuals can recover from livestock losses due to drought. But raiding needs to be understood as a response to political as well as environmental events, especially when these political events lead to the differential access to weapons and to regional and national politics that favor one group over another.

Livestock, Family Formation, and Growth

It is only through livestock that men can marry and thus have claim to their children. Theory in evolutionary or behavioral ecology stresses that the underlying motivations that govern human behavior are related to the differential dissemination of an individual's genes, or a close relative's, into the larger population. One of the basic problems with this theoretical approach is that human behavior is often divorced from the cultural context within which it is embedded. An examination of family formation and growth among the Turkana is impossible without an understanding of the cultural norms that govern social relationships. Children are highly valued by the Turkana, and the management of livestock herds is thought to be impossible without them. Men who do not have children are looked upon as failures; in the words of Angorot, "without children you are nothing more than a dog." A woman's status in society is largely a function of the number of children that she has had. Women with few children are not buried when they die; they are just left in the bush or in the fork of a tree and then the family moves on.

For a man, having children does not just mean biologically producing them; it means having them grow up in your household, contributing to the growth of the family and its herds. This is only possible through marriage and the transfer of livestock through bridewealth. Until bridewealth is paid and the marriage is completed, children belong to the woman's father or elder brother. If children are produced and no marriage takes place, the children are raised as part of the woman's father's or brother's household. There is also a fairly stiff fine, six to ten camels, for impregnating a girl and not marrying her, so "disseminating one's genes" outside of marriage has some severe consequences.

Bridewealth is very high among the Turkana; in fact it is among the highest recorded for any pastoral people. During the study period, 50 large animals (camels and cattle) and 50 to 200 small stock was fairly typical as a bridewealth payment. It is often the case that a man cannot get married until he inherits livestock after his father dies.

When speaking to men about their goals in life, the growth of the family and the growth of the livestock herds are almost always intertwined. Angorot once told me, after he had three wives with the awi and one in Lokori, that he hoped to continue to marry as long as he had enough livestock to pay for the bridewealth. His strategy was to bring a new wife into the awi each time one of his wives became menopausal. I cannot pretend to be an expert on how Turkana women think about this issue. I do know that there is often competition and jealousy among wives. However, I have never met a Turkana woman who felt that a small family with only one wife is the most desirable family arrangement. When I asked Imadio how she felt about Angorot marrying again, she said that she wanted Angorot to continue to marry "as long as his heart is still thumping." She stressed the importance of the relationships among in-laws and the need for many children to defend the family from enemies.

Thus the growth of the herds must be understood as part of an articulated strategy that involves both livestock and people. In order to explore this in more depth, I examined the growth of Angorot's and Lorimet's families in relation to herd growth for the ten and sixteen years for which I have data.

Angorot

Before looking at the growth of Angorot's family I thought that it would be useful to list the numbers and types of livestock that had been transferred in bridewealth since the foundation of this large extended family. I am including Angorot and his brothers as all part of one family. Although this is a somewhat unusual situation for Turkana herd owners, all the brothers feel like they are part of the same family, and Angorot has been very generous in allocating animals to his brothers. This arrangement may change following the death of Angorot's mother, Nakedeli, and when the brothers have sons approaching adulthood, but during the entire time that I spent with Angorot, he and his brothers formed a single large family unit. In table 9.3 I list all the livestock that had been given out in bridewealth from 1971 to 1996.

Since 1971 a total of 91 camels, 131 cattles, 403 small stock, and 5 donkeys left the herd in order to bring wives into the family. In most families these losses would be to some extent offset by animals brought into the family as young women were married, but this was not the case for Angorot. His only living sister, Kochodin, was married in 1987 to Kalinyang, probably the most powerful emeron among the Ngisonyoka. While it might be beneficial to be linked to such a powerful man, Kalinyang did not transfer livestock for bridewealth. Whether this is typical of emerons, or an unusual case, I do not know. From talking to Angorot, I got the impression that he had little choice, that he had to go along with whatever decision Kalinyang made.

Bridewealth may be transferred all at once, as was the case with Imadio, or over an extended number of years involving multiple payments, as was the case with Kole. As a result of these transfers, there has been steady growth in the size of the family: from a single married man and one wife in 1971 to six married men, twelve wives, and sixty children in 1996 (fig. 9.2).

The relationship of the size of the family to livestock population is illustrated in figure 9.3. Here I compared the overall number of livestock, as measured in TLUs, to the human population; I also charted

TABLE 9.3.   Livestock Transferred in Bridewealth for Angorot's Family

	Camels	Cattle	Small Stock	Donkeys	For
1971	25	34	50	1	Imadio
1974	20	35	150	1	Nangiro (Aki's first wife)
1981	3	3	0	0	Kole
1984	0	5	28	0	Kole
1984	23	22	20		Amanam (Akopenyon's first wife)
1986	0	0	30	0	Atwom (Angorot's first wife in Lokori)
1986	10	2	0	0	Akeng (Nakwawi's first wife)
1986	0	7	20	2	Ewoi (Erionga's first wife)
1986	0	2	30	2	Nater (Lokora's first wife)
1987	1	14	0	0	Akwale (Aki's second wife)
1988	9	7	80	1	Kole
1988	0	0	15	0	Eomasil (Angorot's second wife in Lokori)
1994	0	0	0	30	Naribu (Akopenyon's second wife)

Source: Author's original data.

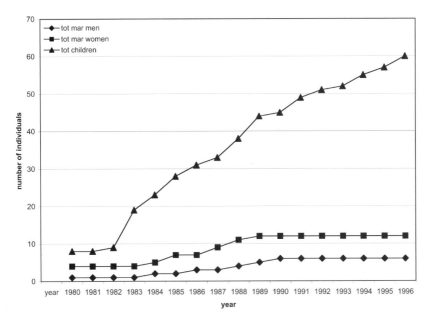

Fig. 9.2.   Angorot: Family growth

the size of the human population. It is clear that the relationship of livestock numbers to family size fluctuated throughout the 1980s, but the family showed a steady increase.

Lorimet

The relationship of overall herd growth to family size is similar for Lorimet, but on a much reduced scale. Lengess returned from Kitale in 1985 and lived together with Lorimet until 1991, when he separated. At this time Lorimet's mother, Natuk, went to live with Lengess. Although Lorimet stressed that the brothers had split amicably, I heard that they argued over the number of livestock allocated to Lorimet's mother, and that this precipitated the decision by Lengess to move away from his brother. The growth of Lorimet's family from the time that I first met him in 1980, when he had one wife and one

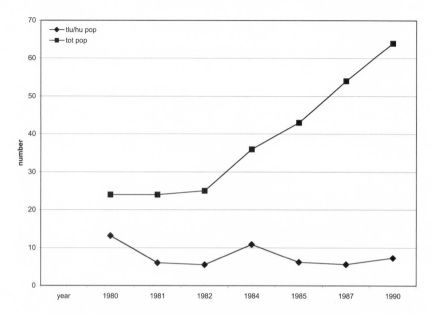

Fig. 9.3.   Angorot: TLUs and human population

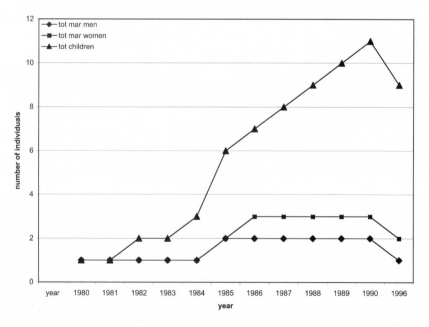

Fig. 9.4.   Lorimet: Family growth

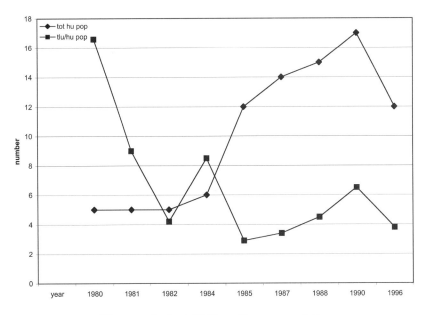

Fig. 9.5.   Lorimet: TLUs and human population

child, to 1996, when he had two wives and nine children, is illustrated in figure 9.4.

An examination of the relationship of livestock numbers to the human population for Lorimet reveals a similar pattern of fluctuating livestock numbers, but a steadily increasing human population as observed for Angorot (fig. 9.5). However, the figure is a bit deceiving. In the early part of the 1980s, especially from 1981 to 1985, the total TLUs exceed that of the TLU/human-population ratio, the only time this was evident in either data set. It needs to be remembered that during this time both the size of Lorimet's family and the number of livestock were very small. It was also a time of severe stress for Lorimet, and he did not have enough livestock to keep him and his family within the pastoral sector without help. Most of the TLUs consisted of cattle, which provide little milk for only a few months each year and for the most part were separated from the major awi. How Lorimet recovered from this stressful period is discussed in the next section.

Recovery and Disaster

Lorimet's Recovery

Following the raid in which Lorimet lost his small stock, he was left with two goats, four large female camels (but of these only one was milking), three nonmilking camels (one was the calf of the milking camel), eight milking cattle with six calves, and four other nonmilking cattle. At this point Lorimet was in danger of being forced out of the pastoral sector. He could not count on the cattle to supply milk throughout the year and had almost no small stock, so the only reliable milk source was a single camel. He joined together with Angorot, his brother-in-law, who was willing to share food with Lorimet and his family. Meanwhile Lorimet and his wife, Asekon, began trying to collect debts, beg livestock from friends and relatives, and conserve what they had left. He began slowly, first collecting a male camel from a man who had killed an ox in the mid-1970s. Another man who had eaten three of Lorimet's sheep was supposed to pay a small ox in return, but Lorimet asked for two female goats and a sheep instead. From in-laws, Lorimet received five female sheep, three female goats, and one donkey, and I gave him enough money to buy five female goats. A daughter of his father's brother got pregnant, and the man responsible had to pay *ekichul* as a penalty. Lorimet received one sheep out of this payment. Lorimet also received three camels and one goat in bridewealth when his half sister, Nangor, was married in 1984. Finally he exchanged a heifer cow for six goats, again in 1984. Through this strategy he was able to rebuild the herd so that by 1985 he had thirty-two cattle, twenty of which were soon to be lost to the Pokot, sixteen camels of which five were milking camels, and a small fence of small stock.

During the period 1985 through 1987 Lorimet continued to build his herd. His brother-in-law, Morukori, gave him a cow, and he was able to get five more cows when his sister Napaipaia was married. He also exchanged a small number of male goats for smaller female ones. By 1987, his small stock flock had grown to 84 animals, and he had 11 camels and 23 cattle. The return of his brother Lengess from Kitale helped tremendously in terms of supplying the labor necessary to manage his small but increasing livestock holdings.

In 1985 he also married for a second time. I have previously described my surprise at this decision and how the old men of the area thought that this was one of Lorimet's best decisions. Lorimet was not able to pay the bridewealth at the time and was indebted to his in-laws, but this demonstrates how decisions related to the growth of herds and that of the family are intertwined.

In 1988, he made his first payment from his herds for the bridewealth of his new wife. He gave seven cattle and three camels to his in-laws, and felt that his herds were growing sufficiently well enough to contribute a male camel to the bridewealth payment of a friend. When I visited him in August 1988, his small stock flock had grown to over 120 animals, and he was herding 31 cattle and 16 camels. I felt that Lorimet had finally recovered by this time. In 1987, he thought he was self-sufficient enough to resume his life as an independent herd owner, moving away from Angorot; by 1988 he was able to begin giving livestock out for social obligations. His herd and flock continued to grow, and by 1990 he was managing 300 small stock, 45 cattle, and 22 camels.

Lopericho and the Dissolution of a Family

The dissolution of Lopericho's family and the decimation of his herds present a starkly different picture from that of Lorimet. Lopericho died in December 1984, probably from a case of cerebral malaria. At the time of his death, he was a very successful herd owner with a growing family and a large holding of livestock based on the raising of camels. Following Lopericho's death, his brother Apuu took over as the principal herd owner and decision maker for the awi. Apuu had been the manager of the cattle herd and remained apart from the rest of the awi for the entire time that I was working with Lopericho. During 1985 the awi members stayed together including Akol, who was married (inherited) to Nawar, Lopericho's full sister. In 1986, Akol separated and moved to the southern side of the Kerio River and then down to Napeitom, dangerously close to the Pokot border. Nawar was afraid to move there and decided to leave the awi and move to the town just north of Lokichar, Lochor Ngikamitak. There she made a living by brewing beer to sell to people who were panning for gold in the area.

Meanwhile Apuu moved to the Naro with the camels and small stock. Unfortunately, this proved to be disastrous as all the camels and most of the small stock were taken in one of the last Pokot raids before they were disarmed later in the same year. With only cattle to rely on, the family was now in deep trouble. Lopericho's wife Lodio and her children moved away and began living with her brother Amodele.

Apuu, his wife and children, and his brothers Komosiong and Iboko all moved to the area near Kaputir and herded their cattle in the gallery forest that is associated with the Turkwell River. Apuu knew the area well and had often brought his cattle herds there in the past. However, it is an area known for frequent outbreaks of livestock disease, and tsetse flies are common in areas of dense bush. They remained there until 1991 when a combination of drought-induced stress and disease hit the herds. Most, if not all, of the cattle died during this dry season, completing the entire destruction of what was once a wealthy and influential family. Apuu moved all the way to the north to the Lokochogio region, close to the Kenya-Sudan border. No one that I talked to knew what he was doing there, but I suspect that he was engaged in raiding the Taposa in an attempt to rebuild his herds. Iboko moved to Kakuma, the site of one of the largest refugee camps in Kenya. Again no one knew what Iboko was doing. In 1996, Komosiong was living near Lokichar with a small flock of goats.

When I last interviewed Nawar in 1996, the family was dispersed and the livestock gone. Turkana herd owners can recover from drought, but they can also see their fortunes turn to dust in a matter of a few hours. Although Lopericho was engaged on a daily basis with herd owners who called the Toma their home, Apuu was not. His decision to move to the Naro and to remain in the Kaputir area after the Pokot raid took him out of the set of relationships that Lopericho had built up over the years. Why Apuu chose this course of actions I do not know.

The knowledge that anyone can lose everything overnight by moving to the wrong place at the wrong time lies in back of each herd owner's mind when he decides to move. It also leads to a sense of fatalism that is often expressed when discussing the future. Frequently when I would be leaving the field for a number of months I would say to people that I would see them when I returned. A frequent response would be, "Yes if I am still alive."

CONCLUSIONS

This chapter has described the livestock dynamics of the individual herd owners that form the basis for this study. I have tried to examine how the demographics of the two populations (livestock and human) are linked. I began the chapter by posing the question, Do Turkana maximize livestock numbers? While I am uncomfortable with terms like *maximize,* increasing herd size was clearly a goal for all the herd owners that I knew—but not just to have more livestock. The data demonstrate the extent to which livestock can buffer the human population from environmental stress. Family size kept increasing even though the overall numbers of livestock fluctuated widely during the study period.

is this smart?

All Turkana herd owners know how vulnerable their livestock are to drought, disease, and raiding. Given that livestock are both a means and an end, exchanging livestock for wives reduces the risk of loss while investing in the growth of the family. Of course a balance needs to be achieved in these decisions. Do the Turkana then maximize the human population? As Leslie has pointed out, there are a number of cultural features within Turkana society that suggest that they do not, such as the late age of marriage, especially for young women. However, if we move away from the thorny discussions of maximization and optimization, then there is no doubt that the growth of the family is extremely important to Turkana men, and that livestock are the means through which this goal is realized.

PART 4

CHAPTER 10

Aggregate and Group Movement

The first rule of the game—to tell the truth—the only rule—
is that the nomad follows the rain.
—R. Capot-Rey, *Le Sahara français*

What can be learned by examining how people use the land and its resources as a group, rather than as individuals? Do groups of Turkana pastoralists merely follow the rain, as suggested in the quote by Capot-Rey? Does the conceptualization of the Turkana environment as a nonequilibrial ecosystem help us understand group behavior and decision making? In this chapter I try to answer these questions by examining Turkana land use at an aggregate level. At the individual level, a herd owner's livestock holdings, family size, and personal characteristics influence his decisions about where and when to move, as well as whether his livestock should be divided into separate herds. Examining data at this fine-grained level of resolution can lead to an emphasis on variability and may mask larger-scale patterns (the phrase "can't see the forest for the trees" comes to mind here). Although the majority of research on pastoral nomads has focused on large-scale patterns, I may have erred in the opposite direction in some of my previous work.

In this chapter I consider the mobility patterns of four other Turkana sections, in addition to that of the Ngisonyoka. I examine how other Turkana sections fared during drought periods and explore the extent to which their land use strategies and the dynamics of their herds correspond to what would be predicted by understanding the environment as a nonequilibrial ecosystem. The incorporation of data from other Turkana sections will also help contexualize the Ngisonyoka material.

The information used here was gathered during a study conducted in 1984 and 1985 as part of a project funded by the Norwegian Agency for International Development (NORAD). The project focused on the impacts and responses to drought in Turkana District and was conducted by ecologists Jim Ellis and David Swift, nutritionist and physical anthropologist Kathleen Galvin, and myself. My role was to understand movement patterns and to collect data on the fluctuations in the livestock population. The complete results of the study are found in our final report, but this has had very limited circulation (Ellis et al. 1987). I have included some aspects of this study in other publications (McCabe 2000), but with a different analytical focus and with less detail than is included here. Although the material may seem outdated, much of it is very similar to the situation in Turkana District today. I will include information gathered from later visits (1986–96) and from recent newspaper stories, but the data collected during the 1984–85 study are still the most detailed yet available.

Group movement among the Turkana typically takes two forms.[1] The first is the movement of people and livestock living together within an adakar. If there is a significant threat of raiding, the members of the adakar will move together, all within a single day. The decisions concerning where to move and on what day are arrived at collectively. Although movement decisions are made collectively, each adakar is commonly associated with one or more wealthy families who wield more influence than others.

The second form is a more common type of group movement. It is the movement of single families and small groups of families and their livestock, all moving in the same general direction within a relatively short period of time. It is not a coordinated movement per se, but taken in the aggregate it represents the movement of a large group of people and their livestock. Decisions here are also arrived at collectively, but concern is with the general direction of movement and the general location where the homesteads will be located. Individual herd owners make their own decisions concerning when to move and where to build their awi. As people move into more dangerous areas, the risk of being raided increases with isolation, so herd owners will rarely choose to be too far out in front of the group or too far behind.

The Turkana sections included in this study are the Ngikamatak, the Ngibocheros, and a combination of two sections (the Ngiluku-mong and the Ngiyapakuno) whose home area is located along the Tarach River in north-central Turkana. The rationale for selecting these sections was that the environments in which they lived had different ecological characteristics from that of the Ngisonyoka, and they appeared to exhibit movement patterns distinct from that of the Ngisonyoka. I am also going to include material from Ngisonyoka, but in the form of aggregate data.[2]

METHODS

This chapter draws on the data relating to group movement and live-stock herd dynamics for which I was responsible in the project described earlier. I will also utilize the results of the data analysis concerning vegetation condition and primary productivity conducted by Jim Ellis and Dave Swift (Ellis et al. 1987) and the livestock census results conducted by Michael Norton-Griffiths and his team at Ecosystems Ltd. (Ecosystems Ltd. 1985). The research on mobility patterns was based on interviews with twenty pastoral families and their livestock from each section during the years 1980 through 1985.[3] This period included two drought episodes (1979–81 and 1984), two "normal years" (1981 and 1982), and two good years (1983 and 1985). Although this climatic pattern was applicable to all the sections included in the study, there were local variations. For instance, the 1984 drought was more severe in the northern than in the southern parts of Turkana District. For each section (except for the Ngibocheros) included in the study we were able to identify a wet season range, a dry season range, and an area that is used primarily during time of drought. Mobility patterns were abstracted from the aggregate data for twenty herd owners in each section. The data were collected by interviewing herd owners with respect to the movements during each dry and wet season for each year. Although people were willing to cooperate with us, I do not trust the data well enough to analyze at the level of individual differences. I believe the data are, in general, accurate and certainly appropriate for the resolution of analysis attempted here.

The data on herd dynamics I found more problematic. I asked people to estimate the number of each species of livestock they had at the beginning and end of each season. I was fully aware that the Turkana do not count their livestock and that any numbers that I was given were gross approximations. Rather than report the actual numbers I calculated an average based on the twenty herd owners in each section. Although I do not trust the accuracy of the initial herd size for each season, there was a remarkable degree of agreement with respect to the approximate changes in herd size during each season.

The estimate of the livestock population for each section is based on work conducted by Ecosystems Ltd., led by Dr. Norton-Griffiths, who is one the world's experts in estimating animal populations from aerial survey. Ecosystems Ltd. was contracted to conduct aerial surveys for all of Turkana District during both the wet and dry seasons in 1984. The estimates for the livestock populations for each section are based on this aerial survey.

Ecologists James Ellis and David Swift of Colorado State University conducted the vegetation analysis. Species composition and biomass productivity were estimated using ground and aerial surveys and through the analysis of remotely sensed imagery using Landsat multispectral scanner (MSS) and very high resolution radiometry (AVHRR). For each of the regions the standing crop of vegetation biomass during an excellent rainfall year was estimated. This figure was used as a baseline so that decrements from this estimate could be used to approximate changes during dry periods and drought years. The data are presented as herbaceous biomass, woody foliage biomass, and dwarf shrub biomass. Ellis cautions that the methods were not as precise as he would have liked, but that they were the same for each region and thus good for comparative purposes (Ellis et al. 1987). I am not so concerned here with the biomass estimates per se but in the comparative framework suggested by Ellis. The areas identified as wet and dry season ranges, as well as drought reserves for each section in the case study, are illustrated in map 10.1.

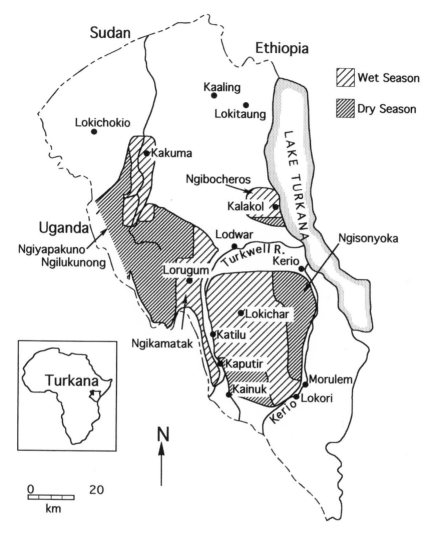

Map 10.1. Wet and dry season ranges for sections

THE CASE STUDY

The Ngikamatak

The Ngikamatak live in west-central Turkana District, and at the time of the study numbered approximately 22,500 people. The topography of this region consists of highlands that merge in to the Rift Valley escarpment, associated footslopes and valleys, and extensive lowlands. The entire area comprises approximately 9,000 square kilometers, of which 2,200 comprise the lowland wet season range and 3,300 the highland dry season range. The remaining 3,500 square kilometers lie above the escarpment in Uganda and include part of the dry season range and the drought reserve range. Precipitation in this area is estimated to vary from approximately 200 to 350 millimeters per year in the lowlands to 600 to 800 millimeters per year in the high elevations. Based on the remote sensing analysis of vegetation Ellis and Swift were able to determine that the dry season range was the most productive area in all of Turkana District, and the wet season range was considered fairly average.

The predominant vegetation type in the Ngikamatak wet season range is bushed grassland, which covers almost 60 percent of the area; the maximum vegetation biomass was estimated at 1,440 kilograms per hectare. A contrasting type is the heavily wooded dry season range, which also includes substantial areas of wooded and bushed grasslands. Total biomass productivity for this range was estimated at 3,060 kilograms per hectare. The dry season and drought reserve ranges in Uganda are assumed to be at least as productive as the dry season range, but we were not able to verify this.

According to the two censuses conducted by Ecosystems Ltd., the average number of livestock in this area was estimated to be 27,400 cattle, 10,500 camels, 174,000 sheep and goats, and 2,200 donkeys. Based on a calculation of maximum vegetation productivity, vegetation offtake by various livestock species, and the livestock population at that time, Ellis and Swift were able to estimate that stocking density of this area was only 13 percent of its maximum potential.

Mobility

The Ngikamatak usually spend the wet season on the plains to the west of Lorigumu (see map 10.2). Many families plant sorghum along

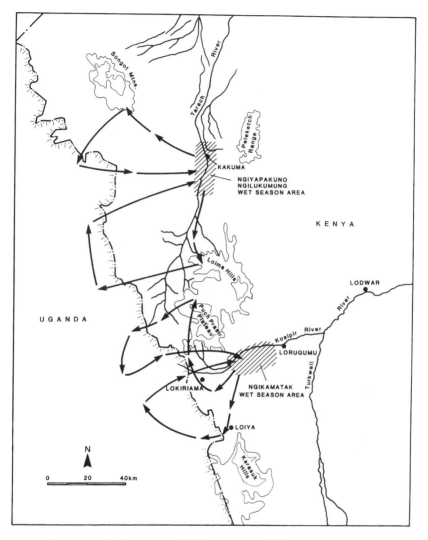

Map 10.2.   Ngiyapakuno, Ngilukumong, and Ngikamatak migration:
Normal year. (From McCabe 2000, in On the Move: How and Why Animals
Travel in Groups, *edited by S. Boinski and P. Garber, p. 670.*
Courtesy the University of Chicago Press.)

the banks of the Kosipir River, and these farms provide an incentive to return to this area each wet season. As the dry season begins, most of the people and the livestock migrate into the dissected foothills along the Rift Valley escarpment. At this time many herd owners divide their livestock, leaving the camels in the foothills, while the cattle, small stock, and most of the family move up into the highlands of the Loima Hills. During even moderately severe dry seasons many of the cattle, small stock, and people then migrate up the escarpment into the rangelands of Uganda. Here they share the range with particular sections (or subtribes) of the Karimojong. The alliance between the Ngikamatak and the Matheniko section of the Karimojong has proved very beneficial to the Ngikamatak, allowing them access to highland pastures. Although this type of alliance is by no means guaranteed, it has persisted for many years.

During periods of drought the Ngikamatak move their livestock far into the Karimojong territory and beyond. During the 1984–85 drought, some of the Ngikamatak herd owners that I interviewed said they brought their cattle to the shores of Lake Kioga in central Uganda. In the drought of 1984, my informants told me that they joined together with many homesteads of Karimojong during this western migration. They stressed the need to move in large groups to defend themselves and their livestock from both the local inhabitants and from the Ugandan army, who frequently attacked their herds in order to obtain meat. Informants also stressed the need to migrate to where the grass was, regardless of the dangers involved. This mobility pattern is illustrated in map 10.3.

Although I am presenting material from the mid-1980s the situation is similar today. In an article that appeared in the *Daily Nation* (December 17, 2000), large numbers of Turkana pastoralists and their livestock were reported to have moved into Kidepo National Park in Uganda. The article went on to mention that these Turkana had joined together with the Matheniko and the Jie to raid the Bakora section of the Karimojong during November 2000. Although no mention was made about the Turkana section that was involved I strongly suspect that it was the Ngikamatak.

Livestock Losses

Despite the extensive movement and the extended period of time above the escarpment in Uganda, the Ngikamatak lost a significant

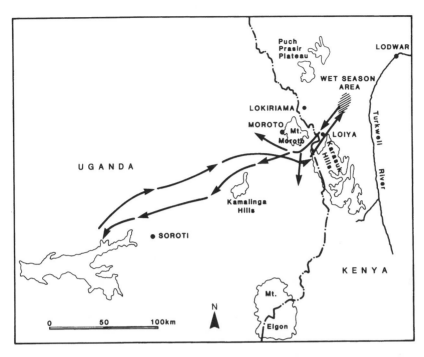

Map 10.3.   Ngikamatak migration: Drought year. (From McCabe 2000, in On the Move: How and Why Animals Travel in Groups, *edited by S. Boinski and P. Garber, p. 672. Courtesy the University of Chicago Press.)*

part of their livestock holdings during this period. The worst period by far occurred during the 1979–80 drought, when 67 percent of the cattle, 48 percent of the camels, and 85 percent of the small stock either died from hunger and disease or were stolen during raids (table 10.1). The major cause of cattle mortality was disease, especially contagious bovine pleuropneumonia (CBPP), which often occurs in conjunction with nutritional stress associated with drought. Other diseases mentioned as important include rinderpest and trypanosomiasis.

Camel mortality was substantially lower than that for cattle during the study period, with the only period of major losses occurring during the 1979–80 dry season. The primary cause was disease, principally viral diarrhea (*cholera*) and hemorrhagic septicemia (*lorogoi*). Other less important diseases were black quarter (*lokichuma*), trypanosomiasis (*edit*), and an unidentified skin disease (*emitina*). The raiding losses were due to Pokot raids in 1981 and 1982 and Tepes raids in 1983 and 1984.

Small stock mortality displayed a similar pattern to that of the cattle and camels. By far the most important factor in small stock mortality was contagious caprine pleuropneumonia (CCPP), with severe outbreaks in the dry seasons of 1979–80 and again in 1980–81. Raiding was the second most important factor in the loss of small stock. Other diseases that caused some mortality were viral diarrhea (referred to as *lotokoyen* in small stock), heart water (*amil*), gastro parasites (*nawosin*), hemorrhagic septicemia, and emitina.

The Tarach Region

The second area considered in the study centers along the Tarach River in northwest Turkana District and is the home area of both the Ngilukumong and the Ngiyapakuno sections of the Turkana. The region is bordered to the south by the Loima and Puch-Prasir plateaus, to the west by the Rift Valley escarpment, and to the north by the Peleketch and Songot mountain ranges. It also includes the southern extension of the Lotikipi plains and the town of Kakuma. At

TABLE 10.1. **Livestock Losses and Causes for Ngikamatak Section, 1980–85**

	Cattle		Camels		Small Stock	
	% Loss	Cause	% Loss	Cause	% Loss	Cause
Dry 1979–80	67	Disease, drought and disease, raiding	48	Disease	85	Drought and disease, disease
Wet 1980	16	Disease	14	Disease	35	Disease
Dry 1980–81	65	Disease, raiding	25	Disease	50	Disease
Wet 1981	17	Disease	1	?	9	Disease
Dry 1981–82	29	Drought, disease, raiding	21	Raiding	26	Disease
Wet 1982	37	Disease, raiding	5	Disease	18	Disease
Dry 1982–83	21	Raiding, disease	1	?	1	Disease
Wet 1983	6	Disease	1	Disease	17	Raiding
Dry 1983–84	9	Disease	14	Raiding	6	Raiding
Wet 1984	18	Raiding	0	—	1	?
Dry 1984–85	39	Ugandan army, disease	10	Disease	2	?
Totals—						
Wet seasons	23		17		25	
Totals—						
Dry seasons	77		83		75	

Source: Author's original data.
Note: Causes are listed in order of importance.

the time of the study, Kakuma was the site of a famine relief camp. Today this camp has been transformed into one of the largest permanent refugee camps in Africa.

The total size of the Tarach region is slightly larger than that of the Ngikamatak and encompasses approximately 9,500 square kilometers. The wet season range consists of the low-lying plains adjacent to the Tarach River and southern Lotikippi plains; it includes approximately 2,400 square kilometers. The dry season range consists of approximately 3,300 square kilometers including Songot and Peleketch mountains and the rugged foothills next to the Ugandan escarpment. The remaining 3,800 square kilometers lies above the escarpment and is considered a drought reserve. However, access to this drought reserve is contentious, at best. Unlike the situation with the Ngikamatak, who have made alliances with sections of the Karamojong that allowed easy access to the high rangelands in Uganda, the Ngiyapakuno and the Ngilukumong have to negotiate and/or fight with the Jie and/or Dodoth each time they migrate up the escarpment. The difficulty associated with access to a drought reserve has resulted in famine conditions more severe than endured by either the Ngikamatak or the Ngisonyoka.

The wet season range receives between 165 and 350 millimeters of precipitation annually. This area, however, is unusual in that the soils are very rich and have a high capacity for water retention. This range also receives considerable runoff from rains falling on the mountains and along the Rift Valley escarpment. Thus the limited rainfall can produce abundant grass in good years. The ecologists estimated the overall biomass production in this region at 2,100 kilograms per hectare but state that they consider this estimate to be low. Bushed and wood grasslands cover about 50 percent of the wet season range, while annual and perennial grasslands comprise 22 percent of the vegetation. The remaining vegetation cover consists of dwarf shrubs and trees.

The dry season range receives from 400 to 800 millimeters of rainfall each year. Bush land and woodland are the dominant vegetation types for this area, and woody canopy covers approximately 46 percent of the dry season range. Primary productivity for the range is estimated to be approximately 3,200 kilograms per hectare.

The drought range receives an equal or greater amount of rainfall as the dry season range, and the biomass productivity should be at least as high as the 3,200 kilograms per hectare estimated for the dry season range. Like the situation with the Ngikamatak this estimate could not be confirmed, as we did not have permission to work in Uganda.

Based on the data from Ecosystems Ltd., the Tarach region held at the time of the study approximately 42,300 cattle, 28,200 camels, 295,000 sheep and goats, and 14,200 donkeys. Ellis and Swift estimated that the livestock density was about 22 percent of its maximum potential (Ellis et al. 1987).

Mobility

The people whose home area is associated with the Tarach River can be divided into two groups: those who have semipermanent settlements and small farms along the river and those whose livelihood is dependent only on livestock (map 10.2). The migratory cycle for those with semipermanent settlements begins in the wet season with all livestock and people together. With the onset of the dry season, cattle are typically moved to the slopes of Songot Mountain, and if conditions are dry, they then move to the foothills along the Rift Valley escarpment. Camels and small stock will usually remain with the major homestead along the Tarach as long as conditions permit and forage is available. During dry periods, herd owners with enough labor often split their camels and small stock into milking and nonmilking herds. The milking animals remain with the awi while the nonmilking move away from the settlement areas but remain along the Tarach. If necessary the nonmilking small stock and camels may move into the foothills of the Peleketch range, but rarely do these herds migrate more than 30 kilometers away from the settlement area.

For those who make their living exclusively from their animals, the wet season is also spent along the Tarach, but they migrate with all their livestock to the west as the dry season begins and remain in the foothills of the Ugandan escarpment as long as forage and water are available. During severe dry seasons camels and some of the small stock will be left behind, while the cattle and some small stock move up the escarpment into Jie territory. This is a dangerous option as relations between these groups have often been quite hostile. Nevertheless, when the environmental conditions deteriorate to the point that remaining below the escarpment means that the livestock could starve, migration into Jie territory becomes a necessity.

During drought the Ngiyapakuno and Ngilukumong do not hesitate to migrate into Uganda. In 1984 they first moved into Jie and Dodoth territories in Uganda, and when forage conditions there deteriorated with the advancing drought, they moved north into the areas controlled by the Taposa and the Didindga in the southern Sudan. During interviews the herd owners that I talked to expressed no fear concerning the movement into these areas or remorse for the local inhabitants. The confrontations along the Ugandan and Sudanese border result in frequent raids and the exchange of large numbers of animals. Unlike the situation with the Ngisonyoka, who were at a distinct disadvantage in terms of access to weapons vis-à-vis the Pokot, the pastoralists in the Tarach region were heavily armed. Once in Uganda, Turkana adakars may remain there months and sometimes years at a time. The cross-boundary raids in this area can be so severe that they threaten relationships between Uganda and Kenya. This occurred in 2000 when the American embassy sent a representative into the region to try to avert an international conflict along the Kenyan, Ugandan, and Sudanese border. The migratory pattern for this group is illustrated in map 10.4.

Livestock Losses
The data for cattle losses during the study period are summarized in table 10.2. Both the wet season of 1980 and the following dry season were disastrous for the cattle herds of the Ngiyapakuno and the Ngilukumong. The cattle were already severely stressed by the wet season of 1980, and the combination of drought-induced nutritional stress combined with disease outbreaks resulted in the death of approximately 70 percent of the herd. Informants did not identify a single disease as particularly bad, but CBPP, rinderpest, trypanosomiasis, and hemorrhagic septicemia were mentioned as contributing to the losses. The dry season of 1980–81 was not much better, but the major problem during this season was raiding by the Jie. I should reiterate that the data reported here are about losses. There is no doubt that the Turkana were raiding as well, but while people were willing to talk to me about their losses, they did not volunteer information about their gains. As the data demonstrate, raiding was a major source of cattle decrements in nearly every year. In addition to the diseases previously mentioned, anthrax (*enomokore*), black quarter (*lokichuma*), and east coast fever (*lokitt*) were all cited as contributing to the losses from disease.

Camel herds did not suffer the degree of loss exhibited by the cat-

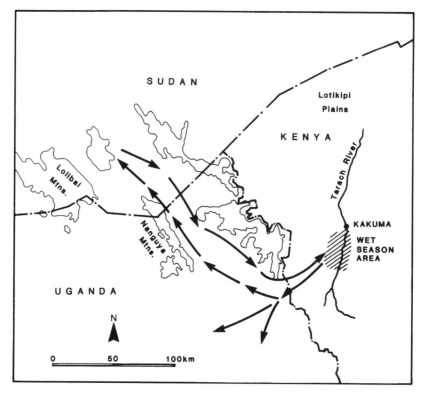

Map 10.4.   Ngiyapakuno migration: Drought year

tle, but two seasons were disastrous. During the wet season of 1980 nearly 70 percent of the camels owned by herd owners in our sample died. The mortality was due to a combination of drought-induced stress and outbreaks of viral diarrhea, trypanosomiasis, east coast fever, anthrax, and hemorrhagic septicemia. The following dry season nearly 70 percent of the remaining camels were lost to Jie and Taposa raids. Although raiding resulted in some losses during six of the ten seasons of the study, large numbers were taken only in 1980–81. Small stock fared better than did either the cattle or the camels for the Ngiyapakuno and the Ngilukumong, with heavy losses occurring only during the wet season of 1980. The drought conditions that began during the 1979–80 dry season intensified, and the small stock that tend to remain close to the awis underwent severe nutritional stress. The only disease that was mentioned as being particularly important was viral diarrhea. No losses were attributed to raiding among the small stock. This was partly due to the fact that many of the goats and sheep tended to remain below the escarpment, but also because they are not as desirable as either species of large stock.

The Ngibocheros

The Ngibocheros represent an atypical section of the Turkana; they incorporate fishing and the hunting of hippopotamus and crocodiles from Lake Turkana and the gathering of wild fruits, along with pastoralism. Although the Ngibocheros as a group engage in all these activities, the pastoralists that I interviewed insisted that they did not engage in any forms of aquatic hunting, nor did they engage in cultivation. They did, however, collect and eat significant amounts of wild foods. They occupy one of the driest areas in Turkana District, bounded on the east by Lake Turkana, to the south by the northern

TABLE 10.2. Livestock Losses and Causes for Ngiyapakuno and Ngilukumong Sections, 1980–85

	Cattle % Loss	Cause	Camels % Loss	Cause	Small Stock % Loss	Cause
Wet 1980	70	Drought and disease, disease, raiding	69	Drought and disease	90	Drought
Dry 1980–81	68	Raiding	18	Raiding	31	Disease
Wet 1981	30	Raiding	20	Disease	1	?
Dry 1981–82	32	Disease, drought, raiding	50	Raiding	2	Disease
Wet 1982	47	Raiding, disease	26	Disease	0	—
Dry 1982–83	25	Drought, raiding	14	Disease, drought	0	—
Wet 1983	16	Disease	30	Raiding, disease	0	—
Dry 1983–84	16	Raiding, disease	18	Disease	0	—
Wet 1984	17	Disease	13	Disease	0	—
Dry 1984–85	15	Disease	3	?	0	—
Totals— Wet seasons	73		72		88	
Totals— Dry seasons	27		229		12	

Source: Author's original data.
Note: Causes are listed in order of importance.

extension of the Kerio Delta, and to the north by the area roughly
defined by the Lodwar-Kalokol road. The topography consists of old
lake basins and dunes with elevations rarely exceeding a hundred
meters above the level of the lake (360 meters). The area occupied by
the Ngibocheros is small compared to other Turkana sections, encom-
passing approximately 1,300 square kilometers. Precipitation
throughout the region is low, averaging approximately 165 millimeters per year. The wet season range of the Ngibocheros includes
approximately 900 square kilometers north of the Turkwell River. The
vegetation here is predominately dwarf shrub grassland and some
bushed and wooded grasslands. The dry season range consists primarily of the 400-square-kilometer area associated with the gallery
forest of the Turkwell River. Herbaceous primary productivity is low
throughout the entire area, averaging 1,000 to 1,200 kilograms per
hectare in an excellent rainfall year. Total primary productivity for the
wet season range was estimated at 1,460 kilograms per hectare, while
that of the dry season range was 1,350 kilograms per hectare. This is
the opposite of what we see in the other sections studied. In all the
other areas the wet season range has the least primary production
while that of the dry season range and the drought reserve has considerably more.

The Ngibocheros are also unusual in that they do not have an area
set aside as a drought reserve. The Ngibocheros keep far fewer livestock than the Ngikamatak, Ngisonyoka, or the people living around
the Tarach, but livestock densities are high. According to the Ecosystems Ltd. census, the livestock in this area consisted of approximately
2,400 cattle, 3,650 camels, 55,600 goats and sheep, and 2,000 donkeys.
Ellis and Swift estimated that the Ngibocheros range supported about
38 percent of its maximum livestock carrying capacity at the time of
the study, as well as about 7,200 people.

Mobility
The Ngibocheros concentrate their livestock management strategies
around the keeping of camels and small stock. Since there are no
mountains nearby and the precipitation is scant, they would have to
engage in long-distance migrations in order to keep large numbers of
cattle. The Ngibocheros seem unwilling to do that.

The pattern of land use is quite different from that of the other sections. Individual families own certain groves of trees (primarily *Acacia*

tortilis) within the Turkwell's gallery forest, while other resources in the area are commonly owned. These groves are located in a family's home area (*ere*), where people and livestock concentrate in the dry season. During the wet season people and livestock move out onto the plains but rarely travel more than 30 kilometers away from their home areas.

This mobility pattern is unique in a number of respects. First, the more typical pattern of returning to one's home area in the wet season is reversed. Second, there is no drought reserve. Third, the Ngibocheros have chosen not to undertake long-distance migrations, regardless of the environmental conditions. When I inquired about this, the most frequent response was that there is no raiding where they are currently living; other more distant areas may have more abundant forage resources, but they are subject to raids. The reason that they are safe from raiding is due to the fact that enemies would have to traverse large expanses of Turkana territory to raid the Ngibocheros. It is possible that a raid could be undertaken given these circumstances, but it is very unlikely that a raiding party could return back to their home with any livestock. The mobility pattern for the Ngibocheros is illustrated in map 10.5.

Livestock Losses
Two aspects of the pattern of losses stand out as distinct from the other sections. First, there were very few losses attributed to raiding, and these were to Ngoroko, not to raiders from other tribes. Second, the 1984–85 drought appeared to be far more severe for the Ngibocheros than the 1979–80 drought. I was repeatedly told that it was the worst in living memory. These losses are reflected in the data summarized in table 10.3.

Nutritional stress associated with the dry season accounted for most of the losses throughout the study period. Cattle numbers had decreased so much that, according to our estimate, by 1985 few families had any cattle at all, and those that did had only a few. No one disease stood out as dominant with respect to cattle mortality, with CBPP, trypanosomiasis, and cattle pox (*etune*) all contributing to the losses.

Camels also suffered devastating losses during the study period, with the dry seasons of 1980–81 and 1984–85 being the most severe. Drought and the combination of drought and disease were the most

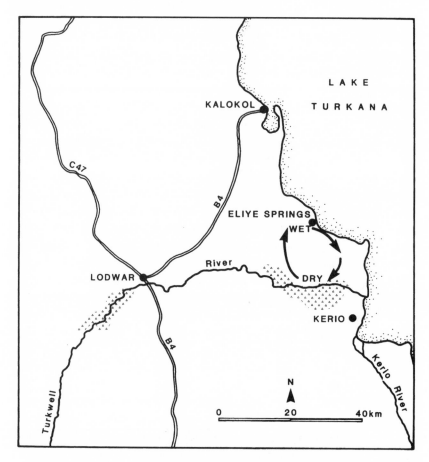

Map 10.5.   Ngibocheros migration: Normal and drought years

important causes for mortality, but trypanosomiasis, pneumonia, hemorrhagic septicemia, and emitina also resulted in camel mortality.

The small stock of the Ngibocheros fared much better than did either the cattle or the camels. However, during the dry seasons of 1980–81 and 1984–85 large numbers of sheep and goats died due to starvation. The people that I interviewed indicated that the numbers of small stock had increased from 1981 to 1984, to the extent that flock size was approximately what it had been before the 1979–80 drought. Unfortunately the Ngibocheros witnessed a 65 percent loss in their small stock the following dry season. While disease was not mentioned as an important factor contributing to small stock mortality (and this may, in fact, have been the case), I suspect that drought and disease both contributed to the loss of goats and sheep. CCPP was a significant factor in all the other areas, and I find it hard to believe that the Ngibocheros area was unaffected. Nevertheless that is what was reported.

The Ngisonyoka

The methodology for the Ngisonyoka component of the study was different from that described for the other Turkana sections. I used the data I had already collected from the four families, supplemented with my knowledge of larger-scale mobility patterns gained from spending over three of the previous five years in the area. I also talked to other herd owners, but not in the interview format used among herd owners from the other sections. These areas will be outlined here (see chap. 4 for a more detailed description).

TABLE 10.3.   Livestock Losses and Causes for Ngibocheros Section, 1980–85

	Cattle % Loss	Cause	Camels % Loss	Cause	Small Stock % Loss	Cause
Wet 1980	21	?	9	Disease	0	—
Dry 1980–81	55	Drought and disease, disease, drought	57	Drought, drought and disease, disease	55	Drought
Wet 1981	3	Disease	6	?	0	—
Dry 1981–82	49	Drought	30	Drought	0	—
Wet 1982	23	Raiding, disease	10	?	0	—
Dry 1982–83	41	Drought	29	Drought, disease	10	Drought
Wet 1983	18	?	18	Disease	0	—
Dry 1983–84	40	Drought	34	Drought	10	?
Wet 1984	21	Disease	14	Disease	0	—
Dry 1984–85	69	Drought	61	Drought	65	Drought
Totals— Wet seasons	27		22		0	
Totals— Dry seasons	73		78		100	

Source: Author's original data.

Note: Causes are listed in order of importance.

The wet season range of the Ngisonyoka encompasses approximately 2,100 square kilometers of low-lying plains located to the east and north of Kailongkol Mountain. The majority of this area is referred to as the Toma and is bordered on the east by Kailongkol and on the west by the Kawaeriwerei Plateau. Annual and perennial grasses are the dominant vegetation type, while woody canopy covers approximately 18 percent of the area. Annual precipitation varies from about 250 to 400 millimeters, and total primary productivity can reach 2,000 kilograms per hectare in very good years.

The Ngisonyoka can be considered to have two dry season ranges. The first encompasses approximately 1,600 square kilometers and is one of the most productive areas in Turkana District. The area consists of the central mountains and the large bushed plains referred to as the Naro, with annual precipitation varying from 400 to 800 millimeters per year. Woody canopy covers approximately 40 percent of the area. Herbaceous vegetation includes both annuals and perennials, and the total primary productivity of this range has been estimated at approximately 3,140 kilograms per hectare in good years.

The southern extension of the dry season range and to some extent the Naro are considered a drought reserve. These areas are usually avoided due to the abundance of ticks, the presence of tsetse, and the risk of raiding. Although not really a drought reserve, the Loriu plateau is an area to which the Ngisonyoka regularly take their cattle in the dry season. It is within the Ngisetu sectional territory, but the Ngisonyoka have a long-standing relationship with members of this section that allows them to move into this area without having to negotiate access.

The second dry season range, which is rarely used, consists of 1,900 square kilometers of dry lava-strewn plains lying to the north and east of the Toma. The area is dry with precipitation ranging from 250 to 400 millimeters. The dominant vegetation types are annual grasses and dwarf shrubs, with total primary productivity averaging 1,400 kilograms per hectare in good years. This area was used only once during the ten years that I closely followed the Ngisonyoka mobility patterns. This occurred during a very good year when the risk of raiding was very high. In normal to dry years this range would not be a viable option, as the forage resources are not sufficient for large numbers of livestock except during wet years.

The remaining land area consists of approximately 1,600 square kilometers of very arid lava plains to the east of the Karweriweri plateau, 800 square kilometers far north of Ngisonyoka territory, and about 800 square kilometers of gallery forest associated with the Turkwell River. The plains to the far north are almost never used due to the area's aridity and lack of water sources. The eastern plains contain numerous salt springs and are used for relatively short periods of time when herd owners feel that their livestock (mostly camels and sometimes goats) need additional salt in their diet. The gallery forest along the Turkwell River was never used during the time that I was studying the Ngisonyoka. The area harbors tsetse flies, the vector for trypanosomiasis, and is very vulnerable to raiding by Pokot. The area also contains many small farms worked by the Ngikaibotok, a group of poor Turkana who have lost their large livestock and subsist through cultivation, beekeeping, and the raising of small livestock, primarily goats.

Based upon the two censuses carried out by Ecosystems Ltd., the livestock holdings of the Ngisonyoka at the time of the study consisted of about 14,000 cattle, 10,800 camels, 124,400 sheep and goats, and 5,000 donkeys. This population was estimated by the ecologists as being about 11 percent of maximum carrying capacity in good years, and supported about 14,500 people at the time of the study.

Mobility
I have already described the mobility patterns for the four Ngisonyoka herd owners in great detail, and a general pattern can also be discerned. The Ngisonyoka spend the wet season in the central plains together with all their livestock. Frequently, individual herd owners will join together into a joint homestead consisting of two to five herd owners in an awi apolon, and these large awis often join together within an adakar. As the dry season sets in, cattle are typically sent either to the foot slopes of the central mountain range or to the Loriu plateau.

As the dry season progresses, the awi tends to move south, and the adakar begins to break up. Frequently the awi apolon will lose one or more herd owners at this time also. If a herd owner has sufficient labor the camels and small stock may be divided into milking and nonmilking herds. Although separate from the major awi, these satellite herds will also be moving south, usually using the rangeland with higher primary productivity located close to the mountains, but farther away from water.

In an average year the Ngisonyoka will only travel as far south as they have to and will return to the Toma shortly after the rains begin. In drought years the people along with the camels and small stock all push to the southern boundaries of the territory, while the cattle will have moved to the pastures on Lataruk and Loretit mountains. These areas are all very dangerous places, close to the border with the Pokot. Once the rains break, the livestock and people typically move into the Naro and then north through the Akalele Pass once it has been determined that there is enough forage in the Toma to support the livestock. This mobility pattern is illustrated in map 10.6.

Herd Dynamics

I have already analyzed and discussed the herd dynamics of the four herd owners, but in order to make this chapter consistent I will briefly summarize the results of this analysis for the five-year period 1980 through 1985.[4] During this time the cattle herds suffered one very bad year in terms of mortality. During the dry season of 1980–81 about 65 percent of the cattle died from a combination of drought and disease. An outbreak of CBPP was the major direct cause of death, but the weakened condition resulting from a lack of forage due to drought was an important contributing factor. In the following dry season, the cattle herd decreased by another 40 percent, but this was due to animals being slaughtered, traded, or sold. The herd increased during the following years and had almost recovered to predrought levels, but in the dry season of 1984–85 the families that I was working with were subject to a series of Pokot raids and lost 83 percent of the remaining cattle (table 10.4).

The camels fared much better than the cattle during this five-year period. They also suffered during the 1980–81 dry season when 41 percent of the herd died due to a combination of drought and disease. There was another outbreak of disease in 1981–82, but then the herds began to recover. By 1985 their numbers had increased to predrought levels. It should be remembered that the 1984–85 drought was not as severe in Ngisonyoka as it was in the central and northern part of Turkana District, and this is reflected in the data on camel mortality.

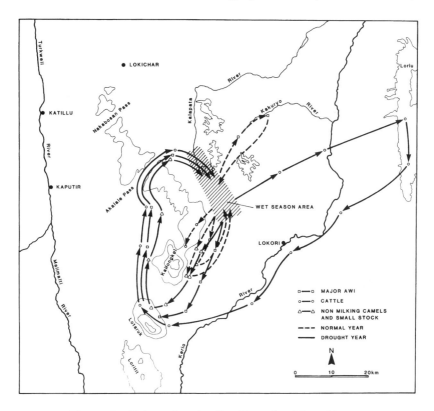

Map 10.6.   Ngisonyoka migration: Normal and drought years.
(From McCabe 1994, in African Pastoralist Systems: An Integrated Approach,
edited by E. Fratkin, K. A. Galvin, and E. A. Roth, p. 80. Copyright © 1994 by
Lynne Rienner Publishers. Used with permission.)

It is also worth mentioning that during the years 1982 through 1985, quite a few camels were traded and given out, either to pay off debts, to help others recover from the drought, or in bridewealth payments.

The Ngisonyoka flocks of goats and sheep suffered a sharp decline during the 1980–81 dry season. Drought-induced famine and the loss of animals in Pokot raids were the principal causes of the losses. In the following wet season there was a severe outbreak of CCPP and many small stock died. Following this very stressful period, the flocks stabilized and then began to rapidly increase. The increase continued until the 1984–85 dry season, when flock size again stabilized.

The Movement of Satellite Herds

The final aspect of group mobility that I want to consider is the movement of single-species satellite herds. When single-species herds are split off from the main awi, the young men and boys who are managing them frequently join together with herders from other satellite herds and live and move together. Although contact is maintained with the main awi, these boys and young men are responsible for the welfare of the herds and flocks, and they make the everyday decisions concerning where and when to move.

Frequently the satellite herds of cattle and small stock will move in similar orbits, but at different times. I have seen this many times among the Ngisonyoka when people and animals were moving south with the dry season. Much has been made in the ecological and rangeland literature about the coordination of different wildlife species in their migratory orbits so that the use of a particular area by one species does not prevent its use by other species, which may be fol-

TABLE 10.4. Livestock Losses and Causes for Ngisonyoka Section, 1980–85

	Cattle % Loss	Cause	Camels % Loss	Cause	Small Stock % Loss	Cause
Wet season 1980 to wet season 1981	65	Disease	41	Drought, disease	55	Famine, raiding
Wet season 1981 to wet season 1982	40	Slaughtered, sold, raided	26	Disease	17	Disease, raiding
Wet season 1982 to wet season 1983	0	—	0	—	0	—
Wet season 1983 to wet season 1984	0	—	0	—	0	—
Wet season 1984 to wet season 1985	83	Raided	0	—	15	Disease

Source: Author's original data.
Note: Causes are listed in order of importance.

lowing behind. Thus species with a narrow range of forage options will precede those animals that are more generalized foragers. This "grazing succession" is also followed among the satellite herds of the Turkana, and in the following vignette the role that the elders play in Turkana forage management is demonstrated.

Dry season 1989: Nadikam, southern Ngisonyoka

I returned to camp in the late afternoon and was told that a couple of herd boys from the small stock camps located on the other side of the Kerio river had been attacked by lions. They had been brought into camp hoping for transport to the hospital, but since I was not there, they were carried down to the shops at Lochorkula looking for help and were able to get a ride into Lokori. A couple of days later I heard that lions had again attacked the small stock camp and some more boys had been injured.

Later that same week I was sitting with a group of men under the "men's tree" when a small group of herd boys approached. They had crossed the river and came to ask for advice from the old men about the recent lion attacks. I was very surprised to see that rather than being sympathetic to the plight of the herd boys, the old men chastised the boys and said that it was their own fault that the lions were attacking. They went on to say that the boys knew well enough that the small stock should not move in front of the cattle on their way south, yet that is just what they were doing. The *ekazicouts* also said that the boys had been warned about doing this and now they were paying for their behavior. The old men issued another warning—the herd boys had better move the goats and sheep in back of the cattle or the attacks would continue. They let it be known that the lions were responding to a curse issued by the old men.

After the boys had left I asked the old men if they had really cursed the herd boys. I had never seen elders deliberately hurt young men or boys and had rarely seen this kind of cursing. They told me that, no, they did not curse the herd boys, but that these boys were just not listening and something had to be done to get them to manage the small stock in the right way.

The "curse" offered the opportunity to get their attention, so they took it. The net result was that the next day the herd boys moved the small stock back to the north and waited until the cattle had moved through before again moving south. The lion attacks stopped.

The Arum-Rum

There is one more aspect of group movement that I want to discuss before concluding this chapter. When I returned to visit friends among the Ngisonyoka in 1996 I was told that many people were living in a new form of social organization that emerged during the last couple of years in response to raiding pressure. This new organization was called an *Arum-Rum* and consisted of many families (perhaps fifty or more) living and moving together within a defensive structure. The Arum-Rum was led by an individual leader, and movement decisions were made by the leader in consultation with an emeron and elder men.

The defensive structure consisted of three concentric rings of thorn fences built one inside the other. Livestock were kept within the innermost fence and let out to pasture and water during the day. At night they were returned to this highly secure place within the Arum-Rum. The people lived outside of the innermost fence but inside the second defensive fence. Families built their own awis there but were in much closer proximity to one another than was usually the case. In between the second fence and the outermost fence small pits were dug and occupied by men with conventional and automatic rifles when people and livestock were in the Arum-Rum.

Once the decision to migrate was made, all household heads were informed, and a coordinated move began on the morning of a designated day. Little dissent was tolerated from individuals or households if they wanted to remain within the Arum-Rum. The Arum-Rum had to move more frequently than individual households as localized resources were exhausted quickly due to the high concentration of livestock.

I was told that the first Arum-Rum was started by Lokorio, a man who was born in Ngisonyoka but who grew up among the Ngikamatak living along the Ugandan border. His grandmother still lived in Ngisonyoka, and when he visited with her in 1994 or 1995 he was shocked by the extent to which the Ngisonyoka were subject to Pokot raids and by their inability to successfully defend themselves. The Ngikamatak regularly live and migrate above the Ugandan escarpment, often in close proximity to enemy groups (sometimes occupying part of their territory). During times when the potential for violence is great they live within an Arum-Rum–type organization. Although depicted as a defensive organization, the Ngikamatak regularly raided out of the Arum-Rum. The political organization based on an overall leader, with movement decisions made in consultation with an emeron in the classic Turkana raiding political hierarchy. It is the only time that the egalitarian ethic is not adhered to.

Lokorio suggested to many Ngisonyoka elders that they form an Arum-Rum in order to defend themselves, and he offered to help. Because the Ngisonyoka did not have enough rifles at the time Lokorio offered to go back to Ngikamatak and return with fighters and their families. He lived up to his promise, and the original Arum-Rum consisted of both Ngisonyoka and Ngikamatak living and moving under the direction of Lokorio.

A second, smaller, Arum-Rum formed within about six months of the first, but this one consisted entirely of Ngisonyoka. Surprisingly, the people with whom I talked said that this Arum-Rum, under the leadership of a man named Ekana, was not well liked, and the leaders of the two Arum-Rums often quarreled over rights to grazing and water. Angorot told me that people from Ekana's Arum-Rum would often take animals from people not living in the Arum-Rum without asking and without paying back the debt incurred. This reminded me of the development of the early Ngoroko groups described earlier in the book.

This large defensive structure was not entirely new to the Ngisonyoka. In the late 1970s one group of Ngisonyoka, called the Aram Family, lived with an organization that sounded very similar to that described as the Arum-Rum. Lopericho and his family lived and traveled with the Aram Family in 1979. The Aram Family was broken up by the GSU (Kenyan army) when they conducted their disarming campaign in 1979. According to Lopericho many men were killed, women raped, and livestock taken during the assault on the Aram Family, a description I heard repeatedly characterizing the entire GSU "operation."

Not all Ngisonyoka chose to live within the Arum-Rums; indeed neither Angorot nor Lorimet ever joined one. However, Angorot's brother Erionga, his family, and the entire herd of cattle did live within the Arum-Rum. What is interesting here in terms of group movement is that a pattern that typifies the regular movement of people and livestock can be altered so quickly and radically. The egalitarian ethic that seems so integral to Turkana social and political organization is abrogated and replaced by a strict hierarchical structure, one that in normal circumstances comes into being for very short periods of time and for a very specific purpose. It is yet another indication of how raiding impacts all aspects of life among the Turkana people—from politics to the daily herding of livestock.

Comparison and Conclusions

Based on the description and data provided here it is clear that there are some patterns in both land use and herd dynamics that are common to most and sometimes all of the Turkana sections included in the study. The exception is the Ngibocheros who manage their livestock in a very different manner than the Ngisonyoka, Ngikamatak, or the people living along the Tarach. For each of the latter groups, their territories consist of a wet season home range, a dry season range, and a drought reserve. The drought reserve is used during times of stress, and its use is often accompanied by significant risk, especially the risk of being raided. For the Ngikamatak, the Ngilukumong, and the Ngiyapakuno, the drought reserve can be considered as located outside of the local ecological and political system (in this case above the escarpment in Uganda). For the Ngisonyoka the same could be said about the Loriu Plateau, which is within the territorial boundaries of the Ngisetu. In each area the wet season range is smaller than the dry season range and of much lower primary productivity. The drought reserve is large and contains the highest biomass productivity within each group's territory (table 10.5).

Although this patterning seems quite logical, the use of areas of low productivity in the wet season and then migration into areas of progressively higher primary productivity with decreasing precipitation has rarely been reported in the literature. It suggests a well-developed strategy of range use that is regular and repeated annually. Of course there is variability, but the overall pattern is there. Other studies have noted the relationship of range productivity to mobility patterns, but in a way that indicates very little in the way of strategic planning. For example, Ekvall has noted that pastoralists in Tibet "follow the growth of grass" (1968:34), and Capot-Rey, in his account of pastoralism in the Sahara, has written that "the first rule of the game—to tell the truth—the only rule—is that the nomad follows the rain" (1953:251). Rephrasing these observations in ecosystem terms, Croze has commented that pastoralists in East Africa "chase ephemeral protein and water around the ecosystem" (Croze and Gwynne 1981:350).

The view that nomadic pastoralists "follow the rain" or have "opportunistic management strategies" suggests a mobility pattern with little regularity or structure. The data and analysis presented here offer a different perspective. Individuals may be seen as moving from one patch of forage resources to another, consistent with an understanding of opportunistic herd management. But this opportunistic movement takes place within broader patterns of land use and livestock management.

The movement of animals to areas of low productivity in the wet season, and then to more productive rangelands during the dry sea-

TABLE 10.5.   Range Characteristics for Four Turkana Sections

Range Characteristics	Ngikamatak	Tarach	Ngibocheros	Ngisonyoka
Total size	9,000 km^2	9,500 km^2	1,300 km^2	8,600 km^2 [a]
Size of wet season range	2,200 km^2	2,400 km^2	900 km^2	2,100 km^2
Productivity of wet season range	1,400 kg/ha	2,100 kg/ha	1,460 kg/ha	2,000 kg/ha
Size of dry season range	3,300 km^2	3,300 km^2	400 km^2	(a) 1,600 km^2 (b) 1,900 km^2 [b]
Productivity of dry season range	3,060 kg/ha	3,200 kg/ha	1,200 kg/ha	(a) 3,140 kg/ha (b) 1,400 kg/ha
Size of drought reserve	3,500 km^2	3,800 km^2	—	Southern part of dry season range
Productivity of drought reserve	3,100+ kg/ha	3,200+ kg/ha	—	3,140+ kg/ha

Source: Data compiled by author.
[a]This includes a 1,600 km area that is almost never used.
[b]Very rarely used

son, resembles that reported for many migratory ungulates. Similar patterns of range use are found among wildebeest in the Serengeti (Pennycuik 1975; Fryxell 1995) and the white-eared cob in the southern Sudan (Fryxell and Sinclair 1988). The explanation for specific migratory movements may vary (about access to forage quantity, see Frixell 1995; about access to forage quality, see Murray 1995), but there is little question that the migration results in an increase in animal numbers compared to what would be possible in a more sedentary system. The same is no doubt true for the Turkana.

One aspect of nonequilibrial ecosystems mentioned by Ellis is that there will be occasions, often associated with drought, when the resources of the system will not be sufficient to sustain the animal population living there (1996). On these occasions, inputs from outside the system are necessary to maintain the population. They may be in the form of imported resources or long-distance migration outside of the system. The mobility patterns described for the Ngika-matak, Ngilukumong, Ngiyapakuno, and to some extent the Ngisonyoka all correspond to this understanding of the functioning of nonequilibrial ecosystems.

The Ngibocheros have obviously adopted a different management strategy. The overall aridity and a large dependence on a single resource, the gallery forest along the Turkwell River, clearly have impacted the way that they manage their livestock. However, the desire to reduce the chance of raiding is the most important variable to be considered in understanding Ngibocheros mobility and decision making. One of the major debates in the literature concerning the movement of animals (wild and domestic) and of people (hunters, foragers, and pastoralists) is the extent to which they "optimize" the use of natural resources (e.g., Deboer and Prins 1989; Smith 1991; Kelly 1995). Central to these debates are strategies intended to maximize access to resources and strategies intended to reduce risk. Couched within this framework, we could say that while the management strategies of the other Turkana sections tend toward increasing access to forage resources, that of the Ngibocheros tends toward reducing risk. I am using the wording *tends toward* deliberately here, as I am uncomfortable saying that the data definitively support a maximization or optimization argument (see also chap. 11). To really assess whether the Turkana are optimizing land use, each movement decision would have to be evaluated with respect to all the alternatives. However, the aggregate data do lend support to an argument for the management strategies of the Ngisonyoka, Ngikamatak, and the Tarach pastoralists as best explained as resource maximization, while that of the Ngibocheros is best explained as risk reduction.

Herd Dynamics

Another key characteristic of nonequilibrial ecosystems is that herbivore populations will experience dramatic losses followed by periods of recovery. The data presented in table 10.6 summarize the losses by species for each of Turkana sections for the two drought periods 1979–81 and 1984–85. The first drought period was clearly the worst for all groups with the exception of the Ngibocheros. Cattle losses ranged from 55 to 70 percent, camel losses ranged from 41 to 69 percent, and small stock losses ranged from 50 to 90 percent. The magnitude of these losses is similar to that reported for other pastoral areas in East Africa, but it should be remembered that this drought lasted for two years.

The impact of the second drought was not as severe, except for the Ngibocheros. For the Tarach pastoralists, the Ngisonyoka, and the Ngikamatak, drought-related losses were high but not devastating. This was not so for the Ngibocheros, who lost around 65 percent of their livestock. The strategy of risk reduction through limited mobility had serious consequences here. The data suggest that highly

TABLE 10.6.   Livestock Losses during Drought (in percentages)

Losses by Species	Ngikamatak	Tarach	Ngibocheros	Ngisonyoka
1980–81				
Cattle	65	70	55	65
	(67% in 1979–80)			
Camels	15	69	57	41
	(48% in 1979–80)			
Small stock	50	90	55	55
	(85% in 1979–80)			
1984–85				
Cattle	39	15	69	83[a]
Camels	10	3	61	0
Small stock	2	0	65	15

Source: Author's original data.
[a]All from raiding.

mobile groups can successfully cope with a single-year drought by utilizing their drought reserve and dividing their livestock but cannot withstand a drought that persists for two years. On the other hand, a more sedentary group like the Ngibocheros is far more vulnerable to drought of a single year's duration. Of course the strategy employed by the Ngisonyoka had serious consequences during 1985 also. They lost a large percentage of their livestock to raids, not drought. I should mention that although the sample size for the Ngisonyoka was small, I know most of the Ngisonyoka cattle were located in same general vicinity as those of Angorot, Lorimet, and Atot during the time of the Pokot raids. These raids involved hundreds of individuals, and everyone I knew lost most of their cattle in this short period of time.

Finally I want to mention that by examining resource use and herd dynamics at the group or aggregate level, the importance of the social and the cultural may be obscured. The concept of territory and rules pertaining to the use of resources (forage and water) within a territory are culturally determined. These territorial boundaries are constantly being contested and negotiated, often in a very violent manner. The previous discussions that focused on individuals emphasized the cultural importance of home areas and social groups; an analysis of herd dynamics is incomplete without considering the importance of social organization to the formation and development of herds and the critical role this plays in recovering from losses. One can only understand family formation and growth as embedded within a social and cultural system. It is my hope that combining the individual- and group-level analyses allows the importance of the ecological and the cultural contexts of land use and decision making to be appreciated.

CHAPTER 11

Conclusions and Discussion

In this final chapter, I return to some of the key issues and questions raised earlier. First, do Turkana mobility and the system of livestock management correspond to what would be expected if the environment in which they live is understood as a persistent but nonequilibrial ecosystem? The second set of questions relate to the possible limitations of this approach. If the Turkana do move and manage their livestock in a way that is consistent with this "new ecological thinking," does this explanation provide sufficient information to understand human-environmental relationships? Are other sorts of ecologies needed? Finally, do the conclusions reached here have policy implications?

TURKANA MOBILITY, LIVESTOCK MANAGEMENT, AND NONEQUILIBRIAL ECOSYSTEMS

It has been hypothesized that herbivore populations living in nonequilibrial ecosystems (1) will be highly mobile, (2) will have opportunistic movement strategies, (3) will display a high degree of variability in movement patterns from one year to the next, and (4) will have highly dynamic populations. It has also been noted that pastoral peoples living in these unstable environments will need to depend occasionally on resources external to the local area and that conflict may be a direct result of the dynamics of this type of ecosystem. I believe the study presented here supports most of these expectations, if not totally, then at least to a large degree.

Mobility, Variability, and Opportunism

The Turkana are one of the world's most mobile peoples, not in the distances moved each year, but in the frequency of their movements. Three out of the four herd owners considered in this study moved their awi more than once a month on average. In dry years the livestock were divided into individual herds of cattle, milking and nonmilking camels, and milking and nonmilking small stock, with the nonmilking animals following a migratory route independent of the awi. By almost any measure, the degree of variability is high. The distances moved and the frequency of moves per year are highly variable for a single herd owner from one year to the next. For any given year, each herd owner followed a migratory pattern that was distinct from those of the other herd owners, sometimes differing radically, suggesting a very different management strategy in response to the same environmental stresses. The variety of ways in which Angorot and Lorimet moved and managed their livestock during drought episodes illustrates this point.

In ecosystems at disequilibrium, mobility patterns should reflect a high degree of opportunism. Opportunistic movements are those that take advantage of temporary patches of resources while avoiding hazards. Opportunistic movement should be reflected in variability on an annual and seasonal basis as well as among herd owners. Variability has been demonstrated, but it was impossible to analyze the data in terms of patch dynamics except as a function of frequency of movement. However, based on many conversations that I had with individual herd owners, I feel confident that herd owners were constantly evaluating different patches of forage resources, taking into account the climate, species mix of their herds, the risk of raiding, and the location of other herd owners and their livestock.

Opportunistic movements are reflected in how the individual herd owners responded to seasonal climatic variations. Each year the Ngisonyoka migrate south, but where they go, the route that they take, and where they stop along the migratory route vary. Nevertheless, as it gets drier, the Ngisonyoka move into areas of progressively higher primary productivity, regardless of their location in any one year.

It is very clear that the herd owners considered in this study place tremendous emphasis on avoiding hazards, to the extent that this is possible. While disease is a major hazard and is taken into account in making a decision where and when to move, the most immediate hazard is raiding. The narrative accounts in chapter 7 illustrate how concerned the individual herd owners were about this threat and the measures they took to avoid being the target of raids. The individual responses relating to the reasons for moving also reflect how important it was for all the herd owners to avoid raids.

For some theorists opportunism allows for the optimization of patch dynamics. Thus Behnke and Scoones remark:

> The producers' strategy within nonequilibrium systems is to move livestock sequentially across a series of environments each of which reaches a peak carrying capacity in a different time period. Mobile herds can then move from zone to zone, region to region, avoiding resource scarce periods and exploiting optimal periods in each area they use. (1993:14)

Although the strategy of mobility outlined by Behnke and Scoones may be largely correct in the abstract, the way this plays out on the ground is far more messy and complicated than can be captured within this type of optimization framework. In the Ngisonyoka case, raids and the threat of raiding greatly influence when and where people and livestock move. At the group level, the Ngisonyoka typically move south as the dry season progresses, but they do not move into the most productive environments except in very dry years. One could argue that the Ngisonyoka are reducing risk rather than maximizing access to forage. That may be true, but all movements are balancing acts, influenced by ecological, social, political, and economic factors. Among other contemporary pastoral peoples, access to markets, schools, medical facilities, and the extent to which they are integrated into regional and national economies exert a significant influence on the way they move and manage their livestock.

Occasional Dependence on External Resources

All of the Turkana sections considered in this account periodically depend on locally unavailable resources. The most obvious example would be the long-distance migrations of the Ngikamatak and the people living along the Tarach River system. In times of drought they often migrate hundreds of kilometers from their home area into areas occupied by other Turkana sections or different ethnic groups. Sometimes these incursions are negotiated and expected, as when the Ngikamatak move above the escarpment and live with the Mathineko section of the Karimojong, or the Ngisonyoka take their cattle to the Loriu plateau, which is within the Ngiesitou section. Other incursions are possible only by force, as when sections from northern Turkana push into the Didinga region in the southern Sudan.

Another example of dependence on external resources is the influx of food aid that was common during the mid-1980s and 1990s. Famine relief is nothing new to Turkana District or other pastoral areas in Africa. In fact, it was so ubiquitous in the 1980s that some governmental and nongovernmental agencies felt that the pastoral livelihood was a remnant of a bygone era and that pastoralism was doomed to failure in today's world.

Conflict

Some authors (e.g., Cousins 1996) argue that conflict is an inevitable outcome of people living in nonequilibrium ecosystems. The argument is underpinned by the need to exploit resources outside of the local ecosystem, and indeed conflict and violence characterize many of the relationships among pastoral groups in northern Kenya, northeastern Uganda, southern Sudan, and southern Ethiopia.

However, access to external resources can be negotiated or fought over. Which path is chosen depends on far more than environmental conditions, and it can only be understood as part of broader political and economic relationships on the regional, state, and international levels.

Highly Dynamic Herbivore Populations and Loose Coupling of Plant-Herbivore Interactions

The livestock population managed by the Turkana is subject to severe crashes, usually brought on by periods of extended drought. This is apparent when analyzing data for individuals as well as for groups (e.g., sections). Losses of 50 to 90 percent are not unusual and occur at least once every ten years. Less severe droughts are common, with a three- to four-year periodicity, and during these times productivity decreases and some livestock die off. These stressful times are punctuated by relatively wet periods where the herds both improve in condition and, depending upon the species of livestock, increase rapidly. It would be hard to find an example of a more dynamic livestock population, and it certainly fits with expectations of a nonequilibrium ecosystem.

The loose coupling of plant-herbivore interactions can be examined by assessing the degree of environmental degradation resulting from numbers of livestock exceeding the carrying capacity of the rangeland. Here I rely on the research conducted by the ecologists working on the South Turkana Ecosystem Project. In an article published in *Science* (1985), Coughenour and colleagues estimated that Ngisonyoka livestock consumed less than 7 percent of the total aboveground net primary productivity and less than 25 percent of the estimated carrying capacity for the Ngisonyoka ecosystem. In a later article, ecologists Ellis and Swift stated: "Despite the dynamic nature of the ecosystem, there is little evidence of degradation or imminent system failure. Instead, this ecosystem and pastoral inhabitants are relatively stable in response to the major stresses on the system, e.g., frequent and severe droughts" (1988:453).

Summary

So where does this leave us? Variability? Absolutely. Opportunistic management? Yes. Occasional dependence on external resources? Yes. Highly dynamic livestock populations and a loose coupling of plant-herbivore interactions? Yes. Do the Turkana system of livestock management and patterns of mobility correspond to what is expected in nonequilibrium ecosystems? Yes. However, does this capture most or all of the human-environmental relationships necessary to understand how the Turkana use the land and its natural resources? No.

OTHER ECOLOGICAL FRAMEWORKS AND THE ENVIRONMENTAL EXPLANATION

Raiding, Violence, and Political Ecology

As an analytic framework, political ecology explains how the use of natural resources at the local level is influenced by the larger political economy within which it is embedded. Harvey (1996) noted that all ecological projects are inherently political-economic projects and that all political-economic projects are inherently ecological. Although not situated within a political ecology framework, Wilmsen's (1989) critique of Richard Lee's (1979) study of the !Kung San in the Kalahari produced a vigorous debate about the importance of understanding how history and political economy combined have shaped local livelihood patterns. Wilmsen's critique is similar to that leveled at many studies in human ecology: that there is a tendency to reduce explanation to ecological relationships while ignoring other political, economic, and institutional contexts that influence access to resources and livelihood strategies.

Stonich's (1993) research has tried to bring together insights from human ecology and political economy in solving real-world problems, while theorists like Escobar (1999) have attempted to problematize the cultural construction of "nature" and bring it into a poststructuralist, antiessentialist political ecology. Some, like Vayda and Walters (1999), feel that the entire political ecology paradigm is flawed in that it privileges political factors on an a priori basis in explaining human-environmental relationships. They criticize some political ecologists who claim to study ecological processes and environmental change for including little in the way of ecological analysis: "It may not be an exaggeration to say that overreaction to the 'ecology without politics' of three decades ago is resulting now in a 'politics without ecology'" (Vayda and Walters 1999:168–69).

In this account I have tried to emphasize the importance of history and politics, while maintaining the ecological underpinnings of the analysis. The historical review demonstrated how raiding of and by the Turkana has been influenced by colonial policy, international relationships, and national politics (chap. 5). More recently, livestock are becoming increasingly commoditized. What formerly was intertribal raiding, where stolen livestock were added to household livestock holdings, has been transformed into an economic enterprise with stolen Turkana livestock being quickly transported to regional and national market centers (the profits accruing to politicians and businessmen).

Raiding and the threat of raiding and violence had great impact on how those I studied managed their livestock and used the land. Each herd owner had to balance the environmental realities of a nonequilibrium ecosystem with political realities of intertribal politics and access to sophisticated weapons in modern-day Kenya. The tension between ecology and politics, as reflected in raiding, became the central theme of this study and in the lives of the Turkana people that I knew. Is this then a study in "political ecology"? I am not sure; it is an attempt to understand the use of the land and its natural resources from more than just an ecological perspective. But I do not believe, as some do, that raiding in this part of the world can be explained as an outcome of the dynamics of nonequilibrial ecosystems. I believe it is crucial to understand raiding within its various historical, economic, and political contexts. In this respect, political ecology is very much a part of this study.

Evolutionary or Behavioral Ecology

It is not uncommon to see theoretical arguments in behavioral ecology set in opposition to theory in ecosystem ecology. In this account decisions made about the transfer of livestock, and the relationship of herd growth to family formation, concern issues more commonly associated with behavioral ecology than ecosystem ecology. But rather than view these theoretical orientations as mutually exclusive, I view them them as complementary and reinforcing.

The Turkana herd owners considered in this book try to increase their herds to the extent possible. They do this within a framework of a nonequilibrial ecosystem. But decisions relating to the management of the livestock herds must take into account a herd owner's desire for a large and growing human family. The periodic crashes of livestock numbers, characteristic of nonequibrium ecosystems, are taken into account when deciding to transfer animals in bridewealth. Livestock then become a means, not just an end.

I have also tried to demonstrate how family formation and growth have to be understood within a cultural context. The transfer of bridewealth legitimates both marriage and fatherhood. Without it a man can neither marry nor claim rights to any children that result from a sexual union. Why do East African pastoralists keep such large herds? At least for the Turkana whom I studied, it is in order to store wealth, provide food and a source of income, and marry more wives and have more children.

Ecosystems and the New Ecological Thinking

Many ecologists and social scientists consider the concept of an ecosystem to be inexorably tied to equilibrium-based feedback relationships (chaps. 1, 2). A challenge to the notion of equilibrium then becomes a challenge to the whole idea of an ecosystem. In contrast, the STEP ecologists came to understand the South Turkana environment as an nonequilibrium ecosystem. Energy still flowed through the system, and nutrients still cycled through the system. The ecosystem, however, was not characterized by stability, but by persistence and resilience. Leslie and colleagues define resilience as "how large a range of conditions is compatible with persistence of the system (Holling 1973), or how quickly a population responds to or recovers from perturbations (Pimm 1991)" (1999:361). The South Turkana ecosystem has persisted for a long period of time, despite wide fluctuations in livestock and human numbers.

Far from exemplifying an ecosystem based on the principles of equilibrium and homoeostatic regulating devices, the Turkana studies show a highly variable and dynamic ecosystem, one that is persistent but not equilibrial, subject to stress by drought and pulsed by short periods of intense precipitation. All organisms living in this type of environment must have strategies for coping with the pressures and uncertainties that characterize it, and that includes humans.

For social scientists like Kottak, Zimmerer, and Scoones the "new ecological thinking" is not about challenging the concept of the ecosystem but about expanding it. The same could also be said about the work of Gunderson and Holling (2002) as well as that of Berkes, Colding, and Folke (2003) working within the "coupled social-ecological systems" framework. They advocate that studies that address the human-environmental relationship must consider concepts like instability, disequilibria, nonlinearity, and complexity, as well as the influence of regional and global forces on local people and places. This study fits well within the parameters of the "new ecological thinking." At its core it is an account of a local group of people constructing their pastoral livelihood in a highly variable and stressful environment that is becoming increasingly dangerous as ethnic differences become politicized, intertribal raiding becomes commoditized, and instability is exacerbated by the influx of modern weapons and regional wars.

The New Ecological Thinking and Its Policy Implications

Proponents of the "new ecological thinking" have advocated a radical shift in development planning for pastoral peoples living in nonequilibrium environments. The basic argument is that for most of the last century development projects targeted at pastoral peoples have been based on the assumption that the environments within which they live are equilibrium-based ecosystems, with tight linkages between the plant communities and herbivore populations. Pastoralists were thought to overstock the range, in accordance with the "tragedy of the commons" thesis, leading to environmental degradation and loss of productivity. The solution to these problems was believed to be some sort of privatized tenure, including individual ranches, group ranches, and grazing blocks, in order to shift tenure systems away from commonly owned and managed rangelands. In addition to privatizing the rangeland, livestock numbers were to be reduced so as not to exceed the "carrying capacity," often through forced destocking programs. The new tenure arrangements required new forms of management and new forms of social organization. These development programs were as much experiments in social engineering as they were programs to increase livestock productivity (McCabe, Schofield, and Perkin 1992). Despite the fact that these types of development projects had an unprecedented history of failure, they persisted throughout most of the twentieth century.

Another major debate in the pastoral development literature has revolved around whether pastoral people are willing to sell livestock if prices are fair. Much of the old thinking postulated that East African pastoral people were overly conservative and unwilling to sell livestock except under the most dire of circumstances. This thinking has its roots in the 1920s, when accounts by early travelers were given additional credence with the publication of Herskovits's 1926 article on the "cattle complex." This was taken as accepted wisdom for much of the twentieth century, much like the idea that pastoral people would degrade their environment by their very nature (Lamprey 1983). During the 1970s, 1980s, and 1990s this notion was repeatedly challenged by many researchers who had extensive experience with pastoral peoples, including myself. The tenor of that debate was that pastoral people were not "conservative" and would willingly sell their livestock with the right incentives. On the one hand, they needed a fair rate of exchange for their livestock. In many, if not most, pastoral areas of eastern Africa the marketing infrastructure has been poor, at best, as well as discriminatory. Traders would offer significantly lower prices for livestock to herders in these areas than they would to people in other, more developed parts of the country. The same traders often charged herders significantly higher prices for grains and other products than in the more developed areas. It is not surprising, then, that pastoralists were unwilling to sell livestock under these inequitable economic circumstances.

On the other hand, and possibly even more important, there was little opportunity to invest economic resources in anything that would provide a greater rate of return than livestock. In many areas of East Africa, not just in pastoral areas, investing in livestock is considered a far better option than putting money in a bank. In most pastoral areas there are no banks, even if pastoralists wanted to sell livestock and deposit their money. In addition, there is usually little opportunity to invest in high-potential land or businesses. A number of research and development studies have tried to address these issues, such as Evangelou's work with the Maasai, Ensminger's study of the transformation of Orma livestock management from a subsistence to a market-oriented system, and the work currently being conducted by Layne Coppock and Peter Little (Little et al. 2001).

The "new" thinking in pastoral development is based on an understanding of arid rangelands (which is where most pastoral people live, especially in Africa) as nonequilibrium ecosystems where people move their livestock extensively and opportunistically. In these types of environments, development projects must be designed to accommodate frequent shifts of people and livestock and allow for occasional migrations outside of the local area in times of stress. In the words of Ellis and Swift, who first brought these development ideas forward, "Reductions in scale or confining pastoralists to ranches is an invitation to disaster" (1988:458). What is being proposed in these views is that the old blueprint prescriptions for pastoral development be replaced by projects that are locally based and incorporate "adaptive management." In Scoones's words, this requires "approaches to planning and intervention that involve adaptive and incremental change based on local conditions and local circumstance" (1996:6). Mobility is to be encouraged, and new flexible forms of pastoral land tenure systems have been advocated.

In addition to understanding the necessity of occasional migrations of pastoral communities to use resources located outside their area, this alternative development paradigm suggests that development planners should take into account the fact that this will at times lead to conflict. Thus development projects must include methods of conflict resolution and must strengthen pastoral institutions with this goal in mind. Planners must expect the occasional crash of livestock numbers and find ways to compensate for the loss of productivity. Improved marketing infrastructure that ensures a free flow of goods is often seen as one intervention that addresses this problem.

Relevance to the Turkana Case Study

The ecological research conducted by members of the South Turkana Ecosystem Project is the basis of much of the new thinking in pastoral development. Therefore, it is not surprising that many of the recommendations would be consistent with Turkana livestock management practices. The research reported in chapter 10 strongly supports the argument that extensive mobility is a critical component in successful adaptation to the nonequilibrium conditions of Turkana District. Those Turkana sections that engaged in long migrations and could tap resources outside of their home ranges were far more successful in coping with and recovering from drought than those whose movements were restricted. The data on Ngisonyoka support the argument that pastoralists move their livestock opportunistically and that management practices vary, both annually and among herd owners.

Despite the correspondence between development paradigms based on an understanding of arid rangelands as nonequilibrium ecosystems and the environment of Turkana District, many of those development alternatives would have little to offer Turkana herders. Before livestock development has any chance of success in Turkana District, the security of the pastoral people and their herds must be assured. This insecurity stems more from political and economic conditions than from the disequilibrial dynamics of the ecosystem. No amount of "adaptive management" will help if the people are killed and their livestock taken in regional ethnic conflicts that are supported, tolerated, or ignored by the state. For the Turkana, this has been going on since the early part of the twentieth century. The same can be said for much of the rangeland occupied by pastoral peoples in Uganda, Ethiopia, Somalia, and the Sudan.

Flexible land tenure systems would help, but this is essentially what different sections of the Turkana have negotiated with many of their neighbors. The Ngikamatak can bring their livestock above the Rift Valley escarpment and mix with the Mathinenko section of the Karimojong, and the Ngisonyoka regularly bring their cattle into Ngiesitou highlands. Here flexible tenure systems would not need to be created, just maintained. The Turkana case study emphasizes the importance of these types of tenure arrangements, but the unstable political realities of that part of the world undermine all development efforts. During the last year the Ugandan army has been trying to disarm the Karimojong and the Pokot living inside of Uganda. The result has been that many heavily armed Pokot have migrated across the border into Kenya, with a corresponding increase in violence and raiding directed at the Turkana. Circumstances well beyond the control of the Turkana, and unrelated to environmental conditions, continue to impact their livelihoods and management practices.

*The new ecological thinking in pastoral development disregards the social aspects of livestock, and in this regard it is lacking. The data and analysis presented here on the relationship between herd growth and family growth add an important dimension to this debate. Since livestock is the means by which families are formed and children legitimated, there is a clear additional disincentive to sell livestock. The transfer of livestock is also important for creating the social ties that are critical to long-term survival in Turkana, where destitution may be only a drought or a raid away. Just as the new ecological thinking must account for social, political, economic, and historical factors in addition to environmental impacts, the embedded political and social contexts of local people's lives must likewise be included to effectively address critical ecological and economic issues in pastoral development.

Summary

My aim in this study has been to bring to the fore the importance of the complex interactions and relationships of social, political, and ecological factors affecting the daily lives and decisions made by a pastoral people living today. I have taken a more inclusive approach to ecology than most, drawing on theoretical perspectives from ecosystem, behavioral, and political ecologies. I find these paradigms more complementary than conflictual, but I still find they can explain only part of the story.

It is essential, therefore, that researchers, development planners, and the audiences they write for (policymakers, governments, nongovernmental organizations) better understand that the implications of these forces, especially for pastoralists in East Africa, are urgent, complex, and far-reaching.

Appendix: South Turkana Ecosystem Project (STEP) Publications

Compiled by Paul W. Leslie and J. Terrence McCabe

Bamberg, I.

1986 Effects of clipping and watering frequency on production of the African dwarf shrub *Indigofera spinosa*. Master's thesis, Colorado State University.

Bartlett, E., and N. Dyson-Hudson, eds.

1980 *The man in the biosphere program in grazing lands.* Proceedings of Sessions Sponsored by the U.S. Man and the Biosphere Program at the First International Rangeland Congress, U.S.A., August 1978. U.S. Man and the Biosphere Program, Washington, DC.

Brainard, J. M.

1981 Herders to farmers: The effects of settlement on the demography of the Turkana population of Kenya. Ph.D. dissertation, State University of New York.

1982 Mating structure of a nomadic pastoral population. *Human Biology* 54 (3): 469–75.

1986 Differential mortality in Turkana agriculturalists and pastoralists. *American Journal of Physical Anthropology* 70 (4): 525–36.

1989 Nutritional status and morbidity on an irrigation project in Turkana District, Kenya. *American Journal of Human Biology* 2 (2): 153–63.

1990 *Health and development in a rural Kenyan community.* New York: Peter Lang.

Campbell, B. C., and P. W. Leslie

1995 Reproductive ecology of human males. *Yearbook of Physical Anthropology* 38:1–26.

Campbell, B. C., P. Leslie, K. Campbell, and L. Kirumbi

1996 Male reproductive function and its contribution to reproductive viability among the Turkana. *American Journal of Physical Anthropology.* Suppl. 22:78.

Campbell, C., P. W. Leslie, M. A. Little, J. Brainard, and M. DeLuca
1999 Settled Turkana. In *Turkana herders of the dry savanna: Ecology and biobehavioral response of nomads to an uncertain environment,* edited by M. A. Little and P. W. Leslie, 334–52. Oxford: Oxford University Press.

Campbell, K. L.
1994 Blood, urine, saliva, and dip sticks: Experiences in Africa, New Guinea, and Boston. *Annals of the New York Academy of Sciences* special issue, *Human reproductive ecology: Interactions of environment, fertility, and behavior,* edited by K. Campbell and J. Woods, 709:312–30.

Campbell, K. L., D. Dookhran, P. Leslie, B. Campbell, M. Little, and C. Kigondu
1996 Turkana female reproductive function: Endocrinology and implications. *American Journal of Physical Anthropology* suppl. 22:78–79.

Campbell, K. L., and J. W. Wood, eds.
1994 Human reproductive ecology: Interactions of environment, fertility, and behavior. *Annals of the New York Academy of Sciences* special issue, *Human reproductive ecology: Interactions of environment, fertility, and behavior,* edited by K. Campbell and J. Woods, 709.

Coppinger, K.
1987 Geomorphic control of distribution of woody vegetation and soil moisture in Turkana District, Kenya. Master's thesis, Colorado State University.

Coppock, D. L.
1985 Feeding ecology, nutrition, and energetics of livestock in a nomadic pastoral ecosystem. Ph.D. dissertation, Colorado State University.

Coppock, D. L., J. E. Ellis, and D. M. Swift
1986 Livestock feeding ecology and resource utilization in a nomadic pastoral ecosystem. *Journal of Applied Ecology* 23:573–83.
1988a Seasonal food habits of livestock in South Turkana, Kenya. *East African Agricultural and Forestry Journal* 52:196–207.
1988b Seasonal patterns of activity, travel, and water intake for livestock in South Turkana, Kenya. *Journal of Arid Environments* 14:319–31.

Coppock, D. L., J. E. Ellis, and S. K. Waweru
1987 A comparative *in vitro* digestion trial using inocula of livestock from South Turkana and Kitale, Kenya. *Journal of Agricultural Science* (Cambridge) 110:61–63.

Coppock, D. L., J. E. Ellis, J. Wienpahl, J. T. McCabe, D. M. Swift, and K. A. Galvin
1982 A review of livestock studies of the South Turkana Ecosystem Project. In *Proceedings, Small Ruminant CRSP Workshop,* 168–72. Nairobi: SR-CRSP.

Coppock, D. L., J. T. McCabe, J. E. Ellis, K. A. Galvin, and D. M. Swift
1985 Traditional tactics of resource exploitation and allocation among nomads in an arid African environment. In *Proceedings of the Inter-*

national Rangeland Development Symposium. Salt Lake City: Society for Range Management.

Coppock, D. L., D. M. Swift, and J. E. Ellis
1986   Seasonal nutritional characteristics of livestock diets in a nomadic pastoral ecosystem. *Journal of Applied Ecology* 23:585–95.

Coppock, D. L., D. M. Swift, J. E. Ellis, and K. Galvin
1986   Seasonal patterns of energy allocation to basal metabolism, activity, and production for livestock in a nomadic pastoral ecosystem. *Journal of Agricultural Science* (Cambridge) 107:357–65.

Coppock, D. L., D. M. Swift, J. E. Ellis, and S. K. Waweru
1988   Seasonal nutritional characteristics of livestock forage in South Turkana, Kenya. *East African Agricultural and Forestry Journal* 52 (3):162–75.

Coughenour, M. B.
1986   Plant-large herbivore interactions in two spatially extensive East African ecosystems. In *Proceedings, IV International Congress of Ecology,* SUNY, Syracuse, New York.
1991a   Dwarf shrub and graminoid responses to clipping, nitrogen, and water: Simplified simulations of biomass and nitrogen dynamics. *Ecology Model* 54:81–110.
1991b   A GIS/RS base modeling approach for a pastoral ecosystem in Kenya. In *Proceedings, Resource Technology 90—Second International Symposium on Advanced Technology in Natural Resources Management* (Conference held in Washington, DC, November 1990). American Society for Photogrammetry and Remote Sensing, Bethesda.
1991c   Spatial components of plant-herbivore interactions in pastoral, ranching, and native ungulate ecosystems. *Journal of Range Management* 44:530–42.
1992   Spatial modeling and landscape characterization of an African pastoral ecosystem: A prototype model and its potential use for monitoring drought. In *Ecological indicators,* vol. 1, edited by D. H. McKenzie, D. E. Hyatt, and V. J. McDonald. London: Elsevier Applied Science.

Coughenour, M. B., D. L. Coppock, J. E. Ellis, and M. Rowland
1990   Herbaceous forage variability in an arid pastoral region of Kenya: Importance of topographic and rainfall gradients. *Journal of Arid Environments* 19:147–59.

Coughenour, M. B., D. L. Coppock, M. Rowland, and J. E. Ellis
1990   Dwarf shrub ecology in Kenya's arid zone: *Indigofera spinosa* as a key forage resource. *Journal of Arid Environments* 18:301–12.

Coughenour, M. B., and J. K. Detling
1986   *Acacia tortilis* seed germination responses to water potential and nutrients. *African Journal of Ecology* 24:203–5.

Coughenour, M. B., and J. E. Ellis
1993   Landscape and climatic control of woody vegetation in a dry tropical ecosystem: Turkana District, Kenya. *Journal of Biogeography* 20:383–98.

Coughenour, M. B., J. E. Ellis, and R. G. Popp
    1990   Morphometric relationships and growth patterns of *Acacia tortilis* and *Acacia refiens* in Southern Turkana, Kenya. *Bulletin of the Torrey Botany Club* 117:8–17.
Coughenour, M. B., J. E. Ellis, D. M. Swift, D. L. Coppock, K. Galvin, J. T. McCabe, and T. C. Hart
    1985   Energy extraction and use in a nomadic pastoral population. *Science* 230 (4726): 619–25.
Curran-Everett, L. S.
    1990   Age, sex, and seasonal differences in the work capacity of nomadic Ngisonyoka Turkana pastoralists. Ph.D. dissertation, State University of New York.
    1994   Accordance between VO_2 max and behavior in Ngisonyoka Turkana. *American Journal of Human Biology* 6:761–71.
Curren, L., and K. Galvin
    1999   Subsistence activity patterns and physical work capacity. In *Turkana herders of the dry savanna: Ecology and biobehavioral response of nomads to an uncertain environment*, edited by M. A. Little and P. W. Leslie, 147–63. Oxford: Oxford University Press.
DeLuca, M.
    1996   Survival analysis of intrauterine mortality in a settled Turkana population. *American Journal of Physical Anthropology*, Suppl. 22:96–97.
    1998   Reproductive ecology and pregnancy loss in a settled Turkana population. Ph.D. dissertation, State University of New York.
DeLuca, M. A., and P. W. Leslie.
    1996   Variation in risk of pregnancy loss. In *The anthropology of pregnancy loss: Comparative studies in miscarriage, still birth, and neonatal death*, edited by R. Cecil, 113–30. Oxford: Berg.
Dick, O., and J. E. Ellis
    1986   A Landsat vegetation biomass map of the Turkana District, northern Kenya. *Proceedings of the Twentieth International Symposium on Remote Sensing of the Environment*, December 1986, Nairobi.
Dyson-Hudson, N.
    1980   Strategies of resource exploitation among East African savanna pastoralists. In *Human ecology in savanna environments*, edited by D. R. Harris, 171–84. London: Academic Press.
    1984   Adaptive resource use strategies of African pastoralists. In *Ecology in Practice*, Part I: *Ecosystem Management*, edited by F. DiCastro, F. W. G. Baker, and M. Hadley, 262–73. Dublin: Tycooly International.
    1985   Pastoral production systems and livestock development projects in East Africa. In *Putting people first: Sociological variables in rural development*, edited by M. Cernea, 157–86. Washington, DC: World Bank.
    1991   Pastoral production systems and livestock development projects in East Africa. In *Putting people first: Sociological variables in rural*

development, 2d ed., edited by M. Cernea, 219–56. Washington, DC: World Bank.

Dyson-Hudson, N., and R. Dyson-Hudson

1982   The structure of East African herds and the future of East African herders. *Development and Change* 13:213–38.

1999   The social organization of resource exploitation. In *Turkana herders of the dry savanna: Ecology and biobehavioral response of nomads to an uncertain environment,* edited by M. A. Little and P. W. Leslie, 69–86. Oxford: Oxford University Press.

Dyson-Hudson, R.

1980   Toward a general theory of pastoralism and social stratification. *Nomadic Peoples* 7:1–17.

1981   Indigenous models of time and space as a key to ecological and anthropological monitoring. In *The future of pastoral people,* edited by J. G. Galaty, D. Aranson, P. C. Salzman, and A. Chouinard, 353–58. Ottawa: International Development Research Center.

1983   Understanding East African pastoralism: An ecosystems approach. In *The keeping of animals: Adaptation and social relations in livestock-producing communities,* edited by R. Berleant-Schiller and E. Shanklin, 1–10. Totowa, NJ: Allanheld and Osmun.

1985   South Turkana herd structures and human/livestock ratios. Report on a survey of 75 South Turkana households: June–August 1982. Appendix 3. In *South Turkana nomadism: Coping with an unpredictably varying environment,* R. Dyson-Hudson and J. T. McCabe, 331–65. HRAFlex Books, FL 17–001 Ethnography Series, Human Relations Area Files, Inc. New Haven, CT.

1988   Ecology of nomadic Turkana pastoralists: A discussion. In *Arid lands today and tomorrow: Proceedings of an international research and development conference,* edited by E. E. Whitehead, C. F. Hutchinson, B. N. Timmermann, and R. C. Varady, 701–3. Boulder: Westview.

1989   Ecological influences on systems of food production and social organization of South Turkana pastoralists. In *Comparative socioecology: The behavioral ecology of humans and other mammals,* edited by V. Standen and R. A. Foley, 165–93. Oxford: Blackwell.

1999   Turkana in time perspective. In *Turkana herders of the dry savanna: Ecology and biobehavioral response of nomads to an uncertain environment,* edited by M. A. Little and P. W. Leslie, 25–40. Oxford: Oxford University Press.

Dyson-Hudson, R., and N. Dyson-Hudson

1980   Nomadic pastoralism. *Annual Review of Anthropology* 9:15–61.

Dyson-Hudson, R., and J. T. McCabe

1982   Final Report to the Norwegian Aid Agency (NORAD) on South Turkana Water Resources and Livestock Movements, Nairobi.

1983   Water resources and livestock movements in South Turkana, Kenya. *Nomadic Peoples* 14:41–46.

1985   *South Turkana nomadism: Coping with an unpredictably varying envi-*

ronment. Ethnography Series FL17–001. New Haven: HRAFlex Books.

Dyson-Hudson, R., and D. Meekers
  1996   The universality of African marriage reconsidered: Evidence from Turkana males. *Ethnology* 35:301–20.
  1999   Migration across ecosystem boundaries. In *Turkana herders of the dry savanna: Ecology and biobehavioral response of nomads to an uncertain environment,* edited by M. A. Little and P. W. Leslie, 315–30. Oxford: Oxford University Press.

Dyson-Hudson, R., D. Meekers, and N. Dyson-Hudson
  1998   Children of the house, children of the dancing ground: Ethnographic and contextual analysis of female marriage. Turkana (Kenya). *Journal of Anthropological Research* 54:19–47.

Ellis, J. E.
  1986   A Landsat vegetation biomass map of the Turkana District, northern Kenya. Presented at the Twentieth International Symposium on Remote Sensing of Environment (December 4–10). Nairobi, Kenya.
  1992   Recent advances in arid land ecology: Relevance to agro-pastoral research in the small ruminant CRSP. In *Sustainable crop-livestock systems for the Bolivian highlands,* edited by Valdivial. Columbia: University of Missouri Press.
  In press Dynamics of dryland ecosystems and livestock development in Africa. In *World agricultural resources in the twenty-first century,* edited by S. Breth. Morrilton, AR: Winrock International.
  In press Wildlife and livestock production systems in sub-Saharan Africa. In *Economic-ecological interactions in sub-Saharan Africa,* edited by N. Stenseth.

Ellis, J. E., D. L. Coppock, J. T. McCabe, K. Galvin, and J. Wienpahl
  1984   Aspects of energy consumption in a pastoral ecosystem: Wood use by the South Turkana. In *Wood, energy, and households: Perspectives on rural Kenya,* edited by C. Barnes, J. Ensminger, and P. O'Keefe, 164–87. Stockholm and Uppsala: Beijer Institute and the Scandinavian Institute of African Studies.

Ellis, J. E., M. B. Coughenour, and D. M. Swift
  1991   Climate variability, ecosystem stability, and the implications for range and livestock development. In the Proceedings of the 1991 International Rangeland Development Symposium, *New concepts in international rangeland development: Theories and applications,* edited by R. P. Cincotta, C. W. Gay, and G. K. Perrier, 1–12. Department of Range Science, Utah State University, Logan.
  1993   Climate variability, ecosystem stability, and the implications for range and livestock development. In *Range ecology at disequilibrium: New models of natural variability and pastoral adaptation in African savannas,* edited by R. H. Behnke Jr., I. Scoones, and C. Kerven, 31–47. London: Overseas Development Institute.

Ellis, J. E., and O. Dick
  1985a Landsat vegetation map of Turkana District, Kenya. NORAD, Oslo.
  1985b The vegetation of Turkana District: A Landsat analysis. NORAD, Oslo.
Ellis, J. E., and K. A. Galvin
  N.d. Climate patterns and land use practices in the dry zones of east and west Africa. Manuscript.
Ellis, J. E., K. Galvin, J. T. McCabe, and D. M. Swift
  1987 Pastoralism and drought in Turkana District, Kenya. A Report to NORAD, Nairobi.
  1988 Pastoralism and drought in Turkana District, Kenya. A report to the Norwegian Aid Agency (NORAD). Bellvue, CO: Development Systems Consultants.
Ellis, J. E., and D. M. Swift
  1988 Stability of African pastoral ecosystems: Alternate paradigms and implications for development. *Journal of Range Management* 41 (6): 450–59.
Fry, P. H., and P. W. Leslie
  1985 Population and demography. In *Turkana District resources survey: Main report,* 4.1–4.12. Turkana Rehabilitation Project and Ministry of Energy and Regional Development, Republic of Kenya, Nairobi.
Fry, P.H., and J. T. McCabe
  1986 A comparison of two survey methods on pastoral Turkana migration patterns and the implications for development planning. *Journal of the Overseas Development Institute* (London) Paper 22b:1–17.
Galvin, K. A.
  1984 Season variation in diet of the nomadic pastoral Turkana. *American Journal of Physical Anthropology* 63:159–60 (abstract).
  1985 Food procurement, diet, activities, and nutrition of Ngisonyoka, Turkana pastoralists in an ecological and social context. Ph.D. dissertation, State University of New York.
  1988 Nutritional status as an indicator of impending food stress. *Disasters* 12 (2): 147–56.
  1992 Nutritional ecology of pastoralists in dry tropical Africa. *American Journal of Human Biology* 4:209–21.
Galvin, K. A., D. L. Coppock, and P. W. Leslie
  1994 Diet, nutrition, and the pastoral strategy. In *African pastoralist systems: An integrated approach,* edited by E. Fratkin, K. A. Galvin, and E. A. Roth, 113–31. Boulder: Lynne Rienner.
Galvin, K., and M. Little
  1999 Dietary intake and nutritional status. In *Turkana herders of the dry savanna: Ecology and biobehavioral response of nomads to an uncertain environment,* edited by M. A. Little and P. W. Leslie, 125–45. Oxford: Oxford University Press.

Galvin, K., and S. K. Waweru
    1987   Variation in the energy and protein content of milk consumed by nomadic pastoralists of north-west Kenya. In *Food and nutrition in Kenya: A historical review,* edited by A. A. J. Jansen, H.T. Horelli, and V. J. Quinn, 129–38. Nairobi: UNICEF and University of Nairobi.

Gray, S. J.
    1988   Growth of settled Turkana schoolchildren. Masters thesis, University of New York, Binghamton.
    1992   Infant care and feeding among nomadic Turkana pastoralists: Implications for child survival and fertility. Ph.D. thesis, State University of New York.
    1994a Comparison of effects of breast-feeding practices on birth-spacing in three societies: Nomadic Turkana, Gainj, and Quechua. *Journal of Biosocial Science* 26:69–90.
    1994b Correlates of dietary intake of lactating women in South Turkana. *American Journal of Human Biology* 6:369–83.
    1995   Correlates of breastfeeding frequency among nomadic pastoralists of Turkana, Kenya: A retrospective study. *American Journal of Physical Anthropology* 98:239–55.
    1996a Ecology of weaning among nomadic Turkana pastoralists of Kenya: Maternal thinking, maternal behavior, and human adaptive strategies. *Human Biology* 68:437–65.
    1996b Infant feeding, infant growth, and infant mortality in Turkana. *American Journal of Physical Anthropology* 98:250–55.
    1998   Butterfat feeding in early infancy in African populations: New hypotheses. *American Journal of Human Biology* 10:163–78.
    1999   Infant care and feeding. In *Turkana herders of the dry savanna: Ecology and biobehavioral response of nomads to an uncertain environment,* edited by M. A. Little and P. W. Leslie, 165–85. Oxford: Oxford University Press.

Gray, S. J., P. W. Leslie, and H. A. Akol
    2002   Uncertain disaster: Environmental instability, colonial policy, and the resilience of East African pastoralist systems. In *The last of the nomadic herders,* edited by W. R. Leonard and M. H. Crawford. Cambridge: Cambridge University Press.

Gray, S. J., and I. L. Pike
    1998   Morbidity, pregnancy outcomes, and fitness costs of sedentarization among pastoralist women in Uganda. *American Journal of Physical Anthropology* Suppl. 26:100.

Johnson, B. R., Jr.
    1985   Longitudinal growth among East African nomadic pastoralists: A baseline study. M.A. thesis, State University of New York.
    1990   Nomadic networks and pastoral strategies: Surviving and exploiting local instability in South Turkana. Ph.D. dissertation, State University of New York.

1999   Social networks and exchange. In *Turkana herders of the dry savanna: Ecology and biobehavioral response of nomads to an uncertain environment*, edited by M. A. Little and P. W. Leslie, 89–106. Oxford: Oxford University Press.

Leslie, P. W., J. R. Bindon, and P. Baker
1984   Caloric requirement of human populations: A model. *Human Ecology* 12:137–62.

Leslie, P. W., and B. C. Campbell
1995   Contrasting fertility patterns among nomadic and sedentary males in Turkana, Kenya. *American Journal of Physical Anthropology* Suppl. 20:135.

Leslie, P. W., K. Campbell, C. Campbell, C. Kigondu, and L. Kirumbi
1999   Fecundity and fertility. In *Turkana herders of the dry savanna: Ecology and biobehavioral response of nomads to an uncertain environment*, edited by M. A. Little and P. W. Leslie, 249–78. Oxford: Oxford University Press.

Leslie, P. W., K. L. Campbell, and M. A. Little
1993   Pregnancy loss in nomadic and settled women in Turkana, Kenya: A prospective study. *Human Biology* 65 (2): 237–54.
1994   Reproductive function in nomadic and settled women of Turkana, Kenya. In *Human reproductive ecology: Interactions of environment, fertility, and behavior*, edited by K. L. Campbell and J. W. Wood. *Annals of the New York Academy of Sciences* 709:218–20.

Leslie, P. W., K. L. Campbell, M. A. Little, and C. S. Kigondu
1996   Evaluation of reproductive function in Turkana women with enzyme immunoassays of urinary hormones in the field. *Human Biology* 68:95–117.

Leslie, P.W., and R. Dyson-Hudson
1999   People and herds. In *Turkana herders of the dry savanna: Ecology and biobehavioral response of nomads to an uncertain environment*, edited by M. A. Little and P. W. Leslie, 233–47. Oxford: Oxford University Press.

Leslie, P. W., R. Dyson-Hudson, and P. Fry
1999   Population replacement and persistence. In *Turkana herders of the dry savanna: Ecology and biobehavioral response of nomads to an uncertain environment*, edited by M. A. Little and P. W. Leslie, 281–301. Oxford: Oxford University Press.

Leslie, P. W., R. Dyson-Hudson, E. Lowoto, and Munyesi
1999   Ngisonyoka event calendar. In *Turkana herders of the dry savanna: Ecology and biobehavioral response of nomads to an uncertain environment*, edited by M. A. Little and P. W. Leslie, 375–78. Oxford: Oxford University Press.

Leslie, P. W., and P. H. Fry
1989   Extreme seasonality of births among nomadic Turkana pastoralists. *American Journal of Physical Anthropology* 79:103–15.

Leslie, P. W., P. H. Fry, K. Galvin, and J. T. McCabe
  1988   Biological, behavioral, and ecological influences on fertility in
          Turkana. In *Arid lands today and tomorrow: Proceedings of an interna-
          tional research and development conference,* edited by E. E. White-
          head, C. F. Hutchinson, B. N. Timmermann, and R. C. Varady,
          705–12. Boulder: Westview.
Leslie, P. W., and T. B. Gage
  1989   Demography and human population biology: Problems and
          progress. In *Human population biology: A transdisciplinary science,*
          edited by M. A. Little and J. Hall, 15–44. Oxford: Oxford Univer-
          sity Press.
Leslie, P. W., M. A. Little, R. Dyson-Hudson, and N. Dyson-Hudson
  1999   Synthesis and lessons. In *Turkana herders of the dry savanna: Ecology
          and biobehavioral response of nomads to an uncertain environment,*
          edited by M. A. Little and P. W. Leslie, 355–73. Oxford: Oxford
          University Press.
Little, M. A.
  1980   Designs for human biology research among savanna pastoralists.
          In *Human ecology in savanna environments,* edited by D. R. Harris,
          479–503. London: Academic Press.
  1985   Multidisciplinary and ecological studies of nomadic Turkana pas-
          toralists. *Biology International* (Paris) 11:11–16.
  1988   Introduction to the symposium: Ecology of nomadic Turkana pas-
          toralists. In *Arid lands today and tomorrow: Proceedings of an interna-
          tional research and development conference,* edited by E. E. White-
          head, C. F. Hutchinson, B. N. Timmermann, and R. C. Varady,
          697–700. Boulder: Westview.
  1989   Human biology of African pastoralists. *Yearbook of Physical Anthro-
          pology* 32:215–47.
  1994   Influences of Turkana pastoralists on dry savanna biodiversity. In
          Man, culture, and biodiversity: Understanding interdependencies,
          edited by G. Hauser, M. A. Little, and D. F. Roberts, 33–41. *Biology
          International* Special Issue 32, International Union of Biological Sci-
          ences (IUBS), Paris.
  1995   Growth and development of Turkana pastoralists. In *Research fron-
          tiers in anthropology: Advances in archaeology and physical anthropol-
          ogy,* edited by P. N. Peregrine, C. R. Ember, and M. Ember. Engle-
          wood Cliffs, NJ: Prentice-Hall.
  1996   The growth of pastoral children. In *Biological anthropology: A syn-
          thetic approach to human evolution,* edited by N. T. Boaz and A. J.
          Almquist, 510–11. Upper Saddle River, NJ: Prentice-Hall.
  1997   Adaptability of African pastoralists. In *Human adaptability: Past,
          present, and future,* edited by S. J. Ulijaszek and R. A. Huss-Ash-
          more, 29–60. Oxford: Oxford University Press.
  2001   Lessons learned from the South Turkana Ecosystem Project. In
          Human ecology in the new millennium, edited by V. Bhasin, V. K.
          Sivastava, and M. K. Bhasin, 137–49. Delhi: Kamala-Raj Enterprises.

In press Ecology and human biology of nomadic Turkana pastoralists. In
The last of the nomadic herders, edited by W. R. Leonard and M. H.
Crawford. Cambridge: Cambridge University Press.

In press Growth of African pastoralist children. In *Cambridge encyclopedia
of human growth and development,* edited by S. J. Ulijaszek, F. E.
Johnston, and M. A. Preece. Cambridge: Cambridge University
Press.

Little, M. A., N. Dyson-Hudson, R. Dyson-Hudson, J. E. Ellis, K. A. Galvin, P.
W. Leslie, and D. M. Swift

1990   Ecosystem approaches in human biology: Their history and a case
study of the South Turkana Ecosystem Project. In *The ecosystem
approach in anthropology: From concept to practice,* edited by E. F.
Moran, 389–434. Ann Arbor: University of Michigan Press.

Little, M. A., N. Dyson-Hudson, R. Dyson-Hudson, J. E. Ellis, and D. M. Swift

1984   Human biology and the development of an ecosystem approach.
In *The ecosystem concept in anthropology,* edited by E. F. Moran,
103–32. Boulder: Westview.

Little, M. A., R. Dyson-Hudson, N. Dyson-Hudson, and N. Winterbauer

1999   Environmental variations in the South Turkana ecosystem. In
*Turkana herders of the dry savanna: Ecology and biobehavioral response
of nomads to an uncertain environment,* edited by M. A. Little and
P. W. Leslie, 317–30. Oxford: Oxford University Press.

Little, M. A., R. Dyson-Hudson, P. W. Leslie, and N. Dyson-Hudson

1999   Framework and theory. In *Turkana herders of the dry savanna: Ecology and biobehavioral response of nomads to an uncertain environment,*
edited by M. A. Little and P. W. Leslie, 3–23. Oxford: Oxford University Press.

Little, M. A., R. Dyson-Hudson, and J. T. McCabe

1999   The ecology of South Turkana. In *Turkana herders of the dry savanna:
Ecology and biobehavioral response of nomads to an uncertain environ-
ment,* edited by M. A. Little and P. W. Leslie, 43–65. Oxford:
Oxford University Press.

Little, M. A., K. Galvin, and P. W. Leslie

1988   Health and energy requirements of nomadic Turkana pastoralists.
In *Coping with uncertainty in food supply,* edited by I. de Garine and
G. A. Harrison, 288–315. Oxford: Oxford University Press.

Little, M. A., K. Galvin, and M. Mugambi

1983   Cross-sectional growth of nomadic Turkana pastoralists. *Human
Biology* 55 (4): 811–30.

Little, M. A., K. Galvin, K. Shelley, B. R. Johnson, and M. Mugambi

1988   Resources, biology, and health of pastoralists. In *Arid lands today
and tomorrow: Proceedings of an international research and development
conference,* edited by E. E. Whitehead, C. F. Hutchinson, B. N. Timmermann, and R. C. Varady, 713–26. Boulder: Westview.

Little, M. A., and S. J. Gray

1990   Growth of young nomadic and settled Turkana children. *Medical
Anthropology Quarterly* 4 (3): 296–314.

Little, M. A., S. J. Gray, and P. W. Leslie

1993   Growth of nomadic and settled Turkana infants of northwest Kenya. *American Journal of Physical Anthropology* 92 (3): 273–89.

Little, M. A., S. Gray, I. Pike, and M. Mugambi

1999   Infant, child, and adolescent growth and physical status. In *Turkana herders of the dry savanna: Ecology and biobehavioral response of nomads to an uncertain environment,* edited by M. A. Little and P. W. Leslie, 187–204. Oxford: Oxford University Press.

Little, M. A., and B. R. Johnson Jr.

1985   Weather conditions in South Turkana, Kenya. In *South Turkana nomadism: Coping with an unpredictably varying environment,* edited by R. Dyson-Hudson and J. T. McCabe, 298–314. Human Relations Area Files, Inc. New Haven: HRAFlex Books Ethnography Series.

1986   Grip strength, muscle fatigue, and body composition in nomadic Turkana pastoralists. *American Journal of Physical Anthropology* 69 (3): 335–44.

1987   Mixed longitudinal growth of Turkana pastoralists. *Human Biology* 59 (4): 695–707.

Little, M. A., and P. W. Leslie, eds.

1990   The South Turkana Ecosystem Project. Report to the government of Kenya, Office of the President. Department of Anthropology, State University of New York, Binghamton, and Natural Resources Ecology Laboratory, Colorado State University, Ft. Collins.

1999   *Turkana herders of the dry savanna: Ecological and biobehavioral response of nomads to an uncertain environment.* Oxford: Oxford University Press.

Little, M. A., P. W. Leslie, and K. L. Campbell

1992   Energy reserves and parity of nomadic and settled Turkana women. *American Journal of Human Biology* 4:729–38.

McCabe, J. T.

1983   Land use among the pastoral Turkana. *Rural Africana* 15/16: 109–26.

1984   Food and the Turkana in Kenya. *Cultural Survival* 8 (1): 48–50.

1985   Livestock management among the Turkana: A social and ecological analysis of herding in an East African population. Ph.D. dissertation, State University of New York.

1987   Variation in livestock production and its impact on social organization. *Research in Economic Anthropology* 8:277–93.

1988   Drought and recovery: Livestock dynamics among the Ngisonyoka Turkana of Kenya. *Human Ecology* 15 (4): 371–89.

1990a   Success and failure: The breakdown of traditional drought coping institutions among the pastoral Turkana of Kenya. *Journal of Asian and African Studies* 25:146–60.

1990b   Turkana pastoralism: A case against the Tragedy of the Commons. *Human Ecology* 18 (1): 81–103.

1991a Livestock development, policy issues, and anthropology in East Africa. In *Anthropology and food policy: Human dimensions of food policy in Africa and Latin America,* edited by D. McMillan, 66–85. Athens: University of Georgia Press.

1991b The Turkana of Kenya. In *Nomads,* edited by P. Carmichael, 67–95. London: Collins and Brown.

1994a The failure to encapsulate: Resistence to the penetration of capitalism by the Turkana of Kenya. In *Pastoralists at the periphery: Herders in a capitalist world,* edited by C. Chang and H. Koster, 309–27. Tucson: University of Arizona Press.

1994b Mobility and land use among African pastoralists: Old conceptual problems and new interpretations. In *African pastoralist systems: An integrated approach,* edited by E. Fratkin, K. A. Galvin, and E. A. Roth, 69–89. Boulder: Lynne Rienner.

1995 Turkana. In *Encyclopedia of world culture,* edited by D. Levinson. New Haven: G. K. Hall-Macmillan Press and the Human Relations Area Files.

1996 Markets, regional economies, and tribal pastoralism: The Ngisonyoka Turkana. In *Tribal and peasant pastoralism,* edited by U. Fabietti and P. Salzman, 251–76. Rome: Ibis Press.

2000 Patterns and process of group movement in human nomadic populations: A case study of the Turkana of northwestern Kenya. In *On the move: How and why animals travel in groups,* edited by S. Boinsky and P. Garber, 649–77. Chicago: University of Chicago Press.

2002 The role of drought among the Turkana of Kenya. In *Culture and catastrophe,* edited by A. Oliver-Smith and S. Hoffman, 213–36. Santa Fe: School of American Research Press.

McCabe, J. T., R. Dyson-Hudson, P. W. Leslie, P. H. Fry, N. Dyson-Hudson, and J. Wienpahl

1988 Movement and migration as pastoral responses to limited and unpredictable resources. In *Arid lands today and tomorrow: Proceedings of an international research and development conference,* edited by E. E. Whitehead, C. F. Hutchinson, B. N. Timmermann, and R. C. Varady, 727–34. Boulder: Westview.

McCabe, J. T., R. Dyson-Hudson, and J. Wienpahl

1999 Nomadic movements. In *Turkana herders of the dry savanna: Ecology and biobehavioral response of nomads to an uncertain environment,* edited by M. A. Little and P. W. Leslie, 109–21. Oxford: Oxford University Press.

McCabe, J. T., and J. E. Ellis

1987 Beating the odds in arid Africa. *Natural History* 96 (1): 32–41.

Mugambi, M., and M. A. Little

1983 Blood pressure in nomadic Turkana pastoralists. *East African Medical Journal* 60 (12): 863–69.

Mugambi, M.

1989 Responses of an African dwarf shrub, *Indigofera spinosa,* to competition, water stress, and defoliation. Ph.D. dissertation, Colorado State University.

Patten, R.

1992   Pattern and process in an arid tropical ecosystem: The landscape and system properties of Ngisonyoka Turkana, north Kenya. Ph.D. dissertation, Colorado State University.

Pike, I. L.

1995   Pregnancy outcomes among nomadic Turkana herders of northwestern Kenya. *American Journal of Physical Anthropology* Suppl. 20:172.

1996   The determinants of pregnancy outcome for nomadic Turkana women of Kenya. Ph.D. dissertation, State University of New York.

Quigley, J. A.

1995   Age parameters and hormonal profiles of Turkana males. M.S. thesis, University of Massachusetts.

Reid, R. S.

1992   Livestock-mediated tree regeneration: Impacts of pastoralists on woodlands in dry tropical Africa. Ph.D. dissertation, Colorado State University.

Reid, R. S., and J. E. Ellis

1995   Impacts of pastoralists on woodlands in South Turkana: Livestock-mediated tree regeneration. *Ecological Applications* 5 (4): 978–92.

Senft, R. L., M. B. Coughenour, D. W. Bailey, L. R. Rittenhouse, O. E. Sala, and D. M. Swift

1987   Large herbivore foraging and ecological hierarchies. *BioScience* 37 (11): 789–99.

Shell-Duncan, B.

1993a   Cell-mediated immunocompetence among nomadic Turkana children. *American Journal of Human Biology* 5:225–35.

1993b   Determinants of infant and child morbidity among nomadic Turkana pastoralists of northwest Kenya. Ph.D. dissertation, Pennsylvania State University.

1994   Child fostering among nomadic Turkana pastoralists: Demographic and health consequences. In *African Pastoralist Systems: An Integrated Approach,* edited by E. Fratkin, K. A. Galvin, and E. A. Roth, 147–64. Boulder: Lynne Rienner.

1995   Impact of seasonal variation in food availability and disease stress on the health status of nomadic Turkana children: A longitudinal analysis of morbidity, immunity, and nutritional status. *American Journal of Human Biology* 7:339–55.

1997   The evaluation of infection and nutritional status as determinants of cellular immunosuppression. *American Journal of Human Biology* 9:381–90.

Shell-Duncan, B., and J. W. Wood

1997 The evaluation of delayed-type hypersensitivity responsiveness and nutritional status as predictors of gastrointestinal and acute respiratory infection: A prospective field study among traditional nomadic Kenyan children. *Journal of Tropical Pediatrics* 43:25–32.

Shell-Duncan, B., K. Shelley, and P. W. Leslie

1999 Health and morbidity: Ethnomedical and epidemiological perspectives. *In Turkana herders of the dry savanna: Ecology and biobehavioral response of nomads to an uncertain environment,* edited by M. A. Little and P. W. Leslie, 207–29. Oxford: Oxford University Press.

Shelley, K.

1985 Medicines for misfortune: Diagnosis and health care among southern Turkana pastoralists of Kenya. Ph.D. dissertation, University of North Carolina.

Swift, D. M.

1983 A simulation model of energy and nitrogen balance for free-ranging ruminants. *Journal of Wildlife Management* 47 (3): 620–45.

1985 A model of heat balance and thermoregulatory water loss for animals in natural environments. Ph.D. dissertation, Colorado State University.

Wienpahl, J.

1984a The role of women and small stock among the Turkana. Ph.D. dissertation, University of Arizona.

1984b Women's roles in livestock production among the Turkana of Kenya. *Research in Economic Anthropology* 6:193–215.

Williams, S.

1985 Turkana ornament and re-presentation: A post-modern application of Mauss' 'The Gift.' M.A. thesis, State University of New York.

Winterbauer, N. L.

1995 Rainfall, drought, and variation in body composition among nomadic Turkana pastoralists in northwest Kenya. M.A. thesis, State University of New York.

Wyant, J. G., and J. E. Ellis

1990 Compositional patterns of riperian woodlands in the Rift Valley of northern Kenya. *Vegetation* 89:23–37.

Wyant, J. G., and R. S. Reid

1992 Notes and records: Determining the age of *Acacia tortilis* with ring counts for South Turkana, Kenya: A preliminary assessment. *African Journal of Ecology* 30:176–80.

Notes

CHAPTER 1

1. Other than individual dissertations, Rada Dyson-Hudson and I wrote the only ethnographic account of the people we studied (R. Dyson-Hudson and McCabe 1985). It was published by the Human Relations Area Files Press and has a very limited distribution.

CHAPTER 2

1. *Level shifting* is the use of data from a few sites to determine macroecological processes over a large area.

2. He advocates using local models to examine the relationship among resources, topographic and orographic features, and individual behavior; regional models to examine change over time and exchange relationships; and global models to conduct research on the human dimension of global environmental change (Moran 1990).

CHAPTER 3

1. Lamphear states that the reason that the Turkana were spared from the rinderpest epizootic was because the Turkana cattle were kept in remote mountain pastures, and the livestock that might have come into contact with the disease, the small stock and camels, were not susceptible to the infection.

2. The Ethiopians also raided the Turkana, but historical records suggest that Turkana-Ethiopian relations were principally based on the trade of cattle and ivory for guns.

3. These groups are often referred to as acephalous societies in the anthropological literature.

CHAPTER 4

1. Some earlier publications have estimated the size of Ngisonyoka territory as considerably smaller than 9,600 square kilometers. I feel confident that the figure used here is correct.

2. The vegetation communities mentioned in text are described here. The descriptions are taken from Pratt and Gwynne 1977:46–47.

Bushland

Land supporting an assemblage of trees and shrubs, often dominated by plants of shrubby habit but with trees always conspicuous, with a single or layered canopy, usually not exceeding 10 meters in height except for occasional emergents, and a total canopy cover of not more than 20 percent. The ground cover is poor and fires infrequent; epiphytes can occur. Bushland thicket is an extreme form where woody plants form a closed stand.

Woodland

Land supporting a stand of trees, up to 20 meters in height, with an open or continuous but not thickly laced canopy, and a canopy cover of not more than 20 percent. Shrubs, if present, contribute less than one-tenth of canopy cover. Grasses and other herbs dominate the ground cover.

Shrubland

Land supporting a stand of shrubs, usually not exceeding 6 meters in height, with a canopy cover of not more than 20 percent. Trees, if present, contribute less than one-tenth of the canopy cover. The ground cover is often poor; fires are usually infrequent.

Grassland

Land dominated by grasses and other herbs, sometimes with widely scattered or grouped trees and shrubs, the canopy cover of which does not exceed 2 percent. Usually subject to periodic burning.

Bush Grassland

Grassland with scattered or grouped trees and shrubs, not necessarily equally represented but both always conspicuous and with a combined canopy cover of less than 20 percent. May be subject to periodic burning.

Tree Grassland or Wooded Grassland

Grassland with scattered or grouped trees, the trees always conspicuous, but having a canopy cover of less than 20 percent. Often subject to periodic burning.

Shrub Grassland

Grassland with scattered or grouped shrubs, the shrubs always conspicuous, but having a canopy cover of less than 20 percent. May be subject to periodic burning.

Dwarf Shrub Grassland

Grassland, often sparse grassland, set with dwarf shrubs not exceeding 70 centimeters in height, sometime with scattered large shrubs or stunted trees. Fires are rare. Dwarf shrub grassland with a ground cover of less than 2 percent also occurs. ("Dwarf" can be applied whenever shrubs are under 70 centimeters and trees less than 2 meters in height.)

3. For more information concerning AVHRR see Tucker et al. 1983, 1985; and Prince and Tucker 1986.

4. The very controversial Turkwell Gorge Dam now controls the river. During the research period described in this book the dam construction began, but it was not completed until the late 1990s.

CHAPTER 5

1. Also see Whitehead 2001.

2. In his 1989 article Vayda reserved his previously held position for an ecological explanation for warfare.

3. This was repeatedly reported to me when I visited Turkana in 1996. The arming of the Pokot has also been reported in the *Economist* article of 1994 and in Hendrickson, Armon, and Mearns 1998.

CHAPTER 6

1. This is not the same person as Erionga, Angorot's full brother who herds the cattle.

CHAPTER 7

1. The word *shield* was used by numerous Ngisonyoka herd owners whom I talked with about some families moving ahead of others in their migration south.

2. The research team at the time consisted of Jan Wienpahl, Layne Coppock, Kathy Galvin, Mike Little, Dave Swift, and Jim Ellis.

3. I say that the Pokot were *temporarily* not a serious threat. Many of the Pokot hid their guns, and others tried to buy new guns as soon after the operation was over. By 1990 they were able to resume raiding.

CHAPTER 8

1. This calculation is based on the number of locations for the entire time period. Using only the data for which I have exact information, Angorot spent 24.29 days in each location.

2. The data for the cattle are not as accurate as they are for the awi or other species. The cattle were often in inaccessible places, and, when they visited the awi, I had to question family members who had accompanied the cattle about their movements.

CHAPTER 10

1. There is a third form, one that is rare and relatively new. It is called an *arum-rum* and is an organizational structure formed in response to heavy raiding pressure. I will discuss this in more detail later in the chapter.

2. The data for the Ngisonyoka used the detailed knowledge of the four study families for the period 1980 through 1985. I supplemented this data

with more general interviews conducted with other Ngisonyoka herd owners and members of their families.

3. During the 1990s the Ngisonyoka began to use the rangelands to the south of Lokori. These areas were avoided during most of the time that I was conducting intensive fieldwork, as they were considered too dangerous.

4. This section is under the heading "Herd Dynamic" rather than "Livestock Losses" because the data set for Ngisonyoka is more detailed and includes gains as well as losses.

Bibliography

Albert, B.
  1989 Yanomami "violence": Inclusive fitness or ethnographer's representation? *Current Anthropology* 30:637–40.
  1990 On Yanomami warfare: Rejoinder. *Current Anthropology* 31: 558–63.
Bassett, T. J.
  1986 Fulani herd movements. *Geographical Review* 76 (3): 233–48.
Baxter, P. T. W.
  1979 Boran age-sets and warfare. In *Warfare among East African herders,* edited by K. Fukui and D. Turton, 69–95. Osaka: Senri Ethnological Series 3, National Museum of Ethnology.
Bebbington, A.
  1996 Movements, modernizations, and markets: Indigenous organizations and agrarian strategies in Ecuador. In *Liberation ecologies: Environment, development, social movements,* edited by R. Peet and M. Watts, 86–109. London and New York: Routledge.
Bebbington A., and T. Perreault
  2000 Social capital, development and access to resources in highland Ecuador. *Economic Geography* 30 (4): 395–418.
Behnke, R.
  1994 *Natural resource management in pastoral Africa.* London: Commonwealth Secretariat.
Behnke, R. H., and I. Scoones
  1992 *Rethinking range ecology: Implications for rangeland management in Africa.* London: Overseas Development Institute and International Institute for Environment and Development.
  1993 Rethinking range ecology: Implications for range management in Africa. In *Range ecology at disequilibrium: New models of natural variability and pastoral adaptation in African savannas,* edited by R. H. Behnke, I. Scoones, and C. Kervin, 153–72. London: Overseas Development Institute and International Institute for Environment and Development, Commonwealth Secretariat.
Behnke, R., I. Scoones, and C. Kervin, eds.
  1993 *Range ecology at disequilibrium: New models of natural variability and pastoral adaptation in African savannas.* London: Overseas Development Institute.

Berkes, F., J. Colding, and C. Folke
2003   *Navigating social-ecological systems.* Cambridge: Cambridge University Press.

Blaikie, P.
1989   The use of natural resources in developing and developed countries. In *A world in crisis? Geographical perspectives*, edited by R. J. Johnston and P. J. Taylor, 125–50. London: Methuen.
1995   Changing environments or changing views? A political ecology for developing countries. *Geography* 80 (3): 203–14.

Blaikie, P., and H. Brookfield
1987   *Land degradation and society.* London: Methuen.

Blick, J. P.
1988   Genocidal warfare in tribal societies as a result of European induced culture conflict. *Man* 23:654–70.

Bollig, M.
1990   Ethnic conflicts in North-West Kenya: Pokot-Turkana raiding, 1969–1984. *Zeitschrift für Ethnologie* 155:73–90.

Borgerhoff-Mulder, M.
1988   Kipsigis bridewealth payments. In *Human reproductive behavior*, edited by L. L. Betzig, M. Borgerhoff-Mulder, and P. W. Turke, 65–82. Cambridge: Cambridge University Press.
1992   Demography of pastoralists: Preliminary data on the Datoga of Tanzania. *Human Ecology* 20:1–23.

Borgerhoff-Mulder, M., P. J. Richerson, and J. Bryan
1996   *Principles of human ecology.* Needham Heights, MA: Simon and Schuster.

Borgerhoff-Mulder, M., and D. Sellen
1994   Pastoralist decision-making: A behavioral ecological perspective. In *African pastoralist systems*, edited by E. Fratkin, E. Roth, and K. Galvin, 205–29. Boulder: Lynne Rienner.

Botkin, D. B.
1990   *Discordant harmonies: A new ecology for the twenty-first century.* Oxford: Oxford University Press.

Boyd, R., and P. J. Richerson
1992   Group selection among alternative evolutionarily stable strategies. *Journal of Theoretical Biology* 145:331–42.

Brainard, J. M.
1981   Herders to farmers: The effects of settlement on the demography of the Turkana population of Kenya. Ph.D. dissertation, State University of New York.
1982   Mating structure of a nomadic pastoral population. *Human Biology* 54 (3): 469–75.

Broch-Due, V.
1999   Remembered cattle, forgotten people: The morality of exchange and the exclusion of the Turkana poor. In *The Poor Are Not Us:*

Poverty and Pastoralism, ed. D. M. Anderson, and V. Broch-Due, 50–88. Oxford: James Currey.

Brown, L. H.

1971 The biology of pastoral man as a factor in conservation. *Biological Conservation* 3 (2): 93–100.

Bryant, R. L.

1992 Political ecology: An emerging research agenda in third-world studies. *Political Geography* 11 (1): 12–36.

1997 Beyond the impasse: The power of political ecology in third world environmental research. *Area* 29 (1): 5–19.

Butzer, K. W.

1990 A human ecosystem framework for archeology. In *The ecosystem approach in anthropology: From concept to practice,* edited by E. Moran, 91–130. Ann Arbor: University of Michigan Press.

Capot-Rey, R.

1953 *Le Sahara français.* Paris: Presses Universitaires de France.

Carneiro, Robert L.

1994 War and peace: Alternating realities in human history. In *Studying war: Anthropological perspectives,* edited by S. P. Reyna and R. E. Downs, 3–27. Amsterdam: Gordon and Breach.

Cashdan, E., ed.

1990 *Risk and uncertainty in tribal and peasant economies.* Boulder: Westview.

Caughley, G., N. Shepherd, and J. Short, eds.

1987 *Kangaroos: Their ecology and management in the sheep rangelands of Australia.* Cambridge: Cambridge University Press.

Chagnon, N.

1968 *Yanomamo: The fierce people.* New York: Holt, Rinehart, and Winston.

1988 Life histories, blood revenge, and warfare in tribal population. *Science* 239:985–92.

1990 Reproductive and somatic conflicts of interest in the genesis of violence and warfare among tribesmen. In *The anthropology of war,* edited by J. Hass, 77–104. Cambridge: Cambridge University Press.

1996 Yanomami warfare: A political history. Ferguson, Brian R. (book review). *American Anthropologist* 98 (3): 670–72.

1997 *Yanomamo.* 4th ed. Fort Worth, TX: Harcourt Brace Jovanovich.

Clements, F. E.

1916 *Plant succession: An analysis for the development of vegetation.* Publication 242. Washington, DC: Carnegie Institution of Washington.

Coppock, D. L.

1985 *Feeding ecology, nutrition, and energetics of livestock in a nomadic pastoral ecosystem.* Ph.D. dissertation, Colorado State University.

1993 Vegetation and pastoral dynamics in the southern Ethiopian rangelands: Implications for theory and management. In *Range ecology at disequilibrium: New models of natural variability and pastoral adaptation in African savannas,* edited by R. Behnke, I. Scoones, and C. Kervin, 42–61. London: Overseas Development Institute.

Coppock, D. L., J. E. Ellis, and D. M. Swift

1986   Livestock feeding ecology and resources utilization in a nomadic pastoral ecosystem. *Journal of Applied Ecology* 23:573–83.

1988a   Seasonal food habits of livestock in South Turkana, Kenya. *East African Agricultural and Forestry Journal* 52:196–207.

1988b   Seasonal patterns of activity, travel, and water intake for livestock in South Turkana, Kenya. *Journal of Arid Environments* 14:319–31.

Coppock, D. L., D. M. Swift, and J. E. Ellis

1986   Seasonal nutritional characteristics of livestock diets in a nomadic pastoral ecosystem. *Journal of Applied Ecology* 23:585–95.

Coppock, D. L., D. M. Swift, J. E. Ellis, and S. K. Waweru

1988   Seasonal nutritional characteristics of livestock forage in South Turkana, Kenya. *East African Agriculture and Forestry Journal* 52 (3): 162–75.

Coronil, F.

2001   Perspectives on Tierney's darkness in El Dorado. *Current Anthropology* 43 (3): 265–66.

Coughenour, M. B.

1991   A GIS/RS base modeling approach for a pastoral ecosystem in Kenya. In *Proceedings, Resource Technology 90: Second International Symposium on Advanced Technology in Natural Resources Management* (Conference held in Washington, DC, November 1990). American Society for Photogrammetry and Remote Sensing, Bethesda.

1992   Spatial modeling and landscape characterization of an African pastoral ecosystem: A prototype model and its potential use for monitoring drought. In *Ecological indicators,* vol. 1, edited by D. H. McKenzie, D. E. Hyatt, and V. J. McDonald. London: Elsevier Applied Science.

Coughenour, M. B., D. L. Coppock, and J. E. Ellis

1990   Herbaceous forage variability in an arid pastoral region of Kenya: Importance of topographic and rainfall gradients. *Journal of Arid Environments* 19:147–59.

Coughenour, M. B., and J. E. Ellis

1993   Landscape and climate control of woody vegetation in a dry tropical ecosystem: Turkana District, Kenya. *Journal of Biogeography* 20:383–98.

Coughenour, M. B., J. E. Ellis, D. M. Swift, D. L. Coppock, K. Galvin, J. T. McCabe, and T. C. Hart

1985   Energy extraction and use in a nomadic pastoral population. *Science* 230 (4726): 619–25.

Cousins, B.
  1996 Conflict management for multiple resource users in pastoral and agro-pastoral contexts. *IDS Bulletin* 27 (3): 41–54.
Cronk, L.
  1989 From hunters to herders: Subsistence change as a reproductive strategy among the Mukogodo. *Current Anthropology* 30 (2): 224–34.
  1995 Is there a role for culture in human behavioral ecology? *Evolution and Human Behavior* 16: (3): 185–205.
Croze, H., and G. M. Gwynne
  1981 A methodology for the inventory and monitoring of pastoral ecosystem processes. In *The future of pastoral peoples*, edited by J. Galaty, D. Aronson, P. Salzman, and A. Chouinard, 340–52. Ottawa: International Development Research Center.
Cullis, A., and A. Pacey
  1992 *A development dialogue: Rainwater harvesting in Turkana.* London: Intermediate Technology, Ltd.
Curran-Everett, L. S.
  1990 Age, sex, and seasonal differences in the work capacity of nomadic Ngisonyoka Turkana pastoralists. Ph.D. dissertation, State University of New York.
Curren, L., and K. Galvin
  1999 Subsistence activity patterns and physical work capacity. In *Turkana herders of the dry savanna: Ecology and biobehavioral response of nomads to an uncertain environment*, edited by M. A. Little and P. W. Leslie, 147–63. Oxford: Oxford University Press.
Dahl, G., and A. Hjort
  1976 *Having herds: pastoral herd growth and household economy.* Stockholm: Stockholm Studies in Social Anthropology.
Daly, M., and M. Wilson
  1988 *Homicide.* New York: Aldine de Gruyter.
Deboer, W. F., and H. H. Prins
  1989 Decisions of cattle herdsmen in Burkina Faso and optimal foraging models. *Human Ecology* 17 (4): 445–64.
Deschler, W. W.
  1965 Native cattle keeping in eastern Africa. In *Man, culture, and animals: The role of animals in human ecological adjustment*, edited by A. Leeds and A. Vayda. Publication 78. Washington, DC: American Association for the Advancement of Science.
Dietz, T.
  1987 Pastoralists in dire straits: Survival strategies and external interventions in a semi-arid region at the Kenya/Uganda border: Western Pokot, 1900–1986. Amsterdam: Netherlands Geographical Studies.

Drury, W. H., and I. C. t. Nisbet
    1973   Succession. *Journal of the Arnold Arboretum* 54 (3):331–68.
Dyson-Hudson, N.
    1966   *Karimojong politics.* Oxford: Clarendon.
    1972   The study of nomads. In *Perspectives on Nomadism,* edited by W. Irons and N. Dyson-Hudson, 2–29. Leiden: Brill.
    1985   Pastoral production systems and livestock development projects in East Africa. In *Putting people first: Sociological variables in rural development,* edited by M. Cernea, 157–86. Washington, DC: World Bank.
    1991   Pastoral production systems and livestock development projects in East Africa. In *Putting people first: Sociological variables in rural development,* 2d ed., edited by M. Cernea, 219–56. Washington, DC: World Bank.
Dyson-Hudson, N., and R. Dyson-Hudson
    1969   Subsistence herding in Uganda. *Scientific American* 220 (2): 76–89.
    1980   Nomadic pastoralism. *Annual Review of Anthropology* 9:15–61.
Dyson-Hudson, R.
    1980   Toward a general theory of pastoralism and social stratification. *Nomadic Peoples* 7:1–17.
    1983   Understanding East African pastoralism: An ecosystems approach. In *The keeping of animals: Adaptation and social relations in livestock-producing communities,* edited by R. Berleant-Schiller and E. Shanklin, 1–10. Totowa, NJ: Allanheld and Osmun.
    1985   South Turkana herd structures and human/livestock ratios. Report on a survey of 75 South Turkana households: June–August 1982. Appendix 3. In *South Turkana nomadism: Coping with an unpredictably varying environment,* R. Dyson-Hudson and J. T. McCabe, 331–65. HRAFlex Books, FL 17–001 Ethnography Series, Human Relations Area Files, Inc. New Haven, CT.
    1989   Ecological influences on systems of food production and social organization of South Turkana pastoralists. In *Comparative socioecology: The behavioral ecology of humans and other mammals,* edited by V. Standen and R. A. Foley, 165–93. Oxford: Blackwell.
Dyson-Hudson, R., and J. T. McCabe
    1985   *South Turkana nomadism: Coping with an unpredictably varying environment.* Ethnography Series FL17–001. New Haven: HRAFlex Books.
Dyson-Hudson, R., and D. Meekers
    1999   Migration across ecosystem boundaries. In *Turkana herders of the dry savanna: Ecology and biobehavioral response of nomads to an uncertain environment,* edited by M. A. Little and P. W. Leslie, 315–30. Oxford: Oxford University Press.
Dyson-Hudson, R., and E. A. Smith
    1978   *Human territoriality: An ecological reassessment.* Khartoum: Khartoum University Press.

Ecosystems Ltd.
    1985   Turkana District resources survey. Report submitted to the Government of Kenya.
Ekvall, R.
    1968   Fields on the hoof: Nexus of Tibetan nomadic pastoralism. New York: Holt, Rinehart, Winston.
Ellen, R.
    1982   Systems and regulation. In *Environment, subsistence, and system: The ecology of small-scale social formations*, 177–203. Cambridge: Cambridge University Press.
Ellis, J. E.
    1996   Climate variability and complex ecosystem dynamics: Implications for pastoral development. In *Living with uncertainty*, edited by Ian Scoones, 37–46. London: Intermediate Technology Publications, Ltd.
Ellis, J. E., and D. L. Coppock
    1985   Vegetation patterns in Ngisonyoka, Turkana. In *South Turkana nomadism: Coping with an unpredictably varying environment*, 315–31. Ethnography Series FL17–001. New Haven: HRAFlex Books.
Ellis, J. E., M. Coughenour, and D. Swift
    1991   Climate variability, ecosystem stability, and the implications for range and livestock development. In *Proceedings of the 1991 International Rangeland Development Symposium, New concepts in international rangeland development: Theories and applications*, edited by R. P. Cincotta, C. W. Gay, and G. K. Perrier, 1–12. Department of Range Science, Utah State University, Logan.
    1993   Climate variability, ecosystem stability, and the implications for range and livestock development. In *Range ecology at disequililibrium: New models of natural variability and pastoral adaptation in Africa savannas*, edited by R. H. Behnke, I. Scoones, and C. Kerven, 31–41. London: Overseas Development Institute.
Ellis, J. E., K. Galvin, J. T. McCabe, and D. Swift
    1987   Pastoralism and drought in Turkana District, Kenya. Final Report to the Norwegian Agency for International Development, Oslo and Nairobi.
Ellis, J. E., and D. Swift
    1988   Stability of African pastoral ecosystems: Alternative paradigms and implications for development. *Journal of Range Management* 41 (6): 450–59.
Ember, M.
    1982   Statistical evidence for an ecological explanation of warfare. *American Anthropologist* 84 (3): 645–49.
Escobar, A.
    1999   After nature: Steps to an anti-essentialist political ecology. *Current Anthropology* 40 (1): 1–30.

Evans-Pritchard, E. E.
   1940   *The Nuer: A description of the modes of livelihood and political institu-tions of a Nilotic people.* Oxford: Oxford University Press.
Feil, D. K.
   1987   *The evolution of highland Papua New Guinea societies.* Cambridge: Cambridge University Press.
Ferguson, R. B.
   1995   *Yanomami warfare: A political history.* Santa Fe: School of American Research Press.
   1996   Lawrence Kelly, *War before civilization. American Anthropologist* 98 (3): 670–72.
   2001   Materialist, cultural, and biological theories on why Yanomami make war. *Anthropological Theory* 1:99–116.
Ferguson, R. B., and N. L. Whitehead
   1992   *War in the tribal zone: Expanding states and indigenous warfare.* Seat-tle: University of Washington Press.
Fleisher, M. L.
   1998   Cattle raiding and its correlates: The cultural-ecological conse-quences of market-oriented cattle raiding among the Kuria of Tan-zania. *Human Ecology* 26 (4): 547–72.
   1999   Cattle raiding and household demography among the Kuria of Tanzania. *Africa* 69 (2): 238–57.
Franke, R. W., and B. H. Chasin
   1980   *Seeds of famine.* Monclair, NJ: Allenheld, Osmun.
Fratkin, E.
   1991   *Surviving drought and development: Ariaal pastoralists of northern Kenya.* Boulder: Westview.
Fratkin, E., K. Galvin, and E. A. Roth
   1994   *African pastoral systems: An integrated approach.* Boulder: Lynne Rienner.
Fratkin, E., and E. A. Roth
   1990   Drought and economic differentiation among Ariaal pastoralists of Kenya. *Human Ecology* 18:385–402.
Fried, M. H.
   1975   *The Notion of Tribe.* Menlo Park: Cummings.
Friedman, J.
   1974   Marxism, structuralism, and vulgar materialism. *Man* 9 (3): 444–69.
Fry, D., and K. Bjorkvist, eds.
   1997   *Cultural variation in conflict resolution: Alternatives to violence.* Mah-wah, NJ: Lawrence Erlbaum.
Fryxell, J. M.
   1995   Aggregation and migration by grazing ungulates in relation to resources and predators. In *Serengeti II: Dynamics, management, and conservation of an ecosystem,* edited by A. R. E. Sinclair and P. Arcese, 257–73. Chicago: University of Chicago Press.

Fryxell, J. M., and A. R. E. Sinclair
  1988 Seasonal migration of the white eared kob in relation to resources. *African Journal of Ecology* 26:17–31.
Fukui, K.
  1979 Cattle colour symbolism and inter-tribal homicide among the Bodi. In *Warfare among East African herders,* edited by K. Fukui and D. Turton, 147–77. Osaka: Senri Ethnological Series 3, National Museum of Ethnology.
Fukui, K., and J. Markakis, eds.
  1994 *Ethnicity and conflict in the Horn of Africa.* London: James Currey.
Fukui, K., and J. Markakis
  1994 Introduction. In *Ethnicity and conflict in the Horn of Africa,* edited by K. Fukui and J. Markakis, 1–11. London: James Currey.
Fukui, K., and D. Turton
  1979a Introduction. In *Warfare among East African herders,* edited by K. Fukui and D. Turton, 1–13. Osaka: Senri Ethnological Series 3, National Museum of Ethnology.
  1979b *Warfare among East African herders,* edited by K. Fukui and D. Turton. Osaka: Senri Ethnological Series 3, National Museum of Ethnology.
Galvin, K. A.
  1985 Food procurement, diet, activities, and nutrition of Ngisonyoka, Turkana pastoralists in an ecological and social context. Ph.D. dissertation, State University of New York.
  1988 Nutritional status as an indicator of impending food stress. *Disasters* 12 (2): 147–56.
Galvin, K., and M. Little
  1999 Dietary intake and nutritional status. In *Turkana herders of the dry savanna: Ecology and biobehavioral response of nomads to an uncertain environment,* edited by M. A. Little and P. W. Leslie, 125–45. Oxford: Oxford University Press.
Geertz, C.
  1963 *Agricultural involution.* Berkeley: University of California Press.
Golly, F.
  1984 Historical origins of the ecosystem concept in biology. In *The ecosystem concept in anthropology,* edited by E. Moran, 33–49. Washington, DC: AAAS.
Gray, S. J.
  1992 Infant care and feeding among nomadic Turkana pastoralists: Implications for child survival and fertility. Ph.D. thesis, State University of New York.
  1994 Correlates of dietary intake of lactating women in South Turkana. *American Journal of Human Biology* 6:369–83.
  1998 Butterfat feeding in early infancy in African populations: New hypotheses. *American Journal of Human Biology* 10:163–78.

2000   A memory of loss: Ecological politics, local history, and the evolution of violence. *Human Organization* 59 (4): 401–18.

Gray, S., M. Little, P. Leslie, M. Sundal, and Wiebusch
2003   Cattle raiding, cultural survival, and adaptability of East African pastoralists. *Current Anthropology* 44 (55): 3–30.

Grayzel, J.
1990   Markets and migration: A Fulbe pastoral system in Mali. In *The world of pastoralism: Herding systems in comparative perspective*, edited by J. Galaty and D. Johnson, 35–67. New York: Guilford Press.

Gulliver, P. H.
1951   *A preliminary survey of the Turkana*. New Series 26. Capetown: Commonwealth School of African Studies.

1955   *The family herds: A study of two pastoral tribes*. London: Routledge and Kegan Paul.

1975   Nomadic movements: Causes and implications. In *Pastoralism in tropical Africa*, edited by T. Monod, 369–86. Oxford: Oxford University Press.

Gunderson, L. H., and C. S. Holling.
2002   *Panarchy: Understanding transformations in human and natural systems*. Washington, DC: Island Press

Hallpike, C. R.
1973   Functionalist interpretations of primitive warfare. *Man* 8:451–70.

Hardin, G.
1968   The tragedy of the commons. *Science* 162:1243–48.

Harris, M.
1984a   Animal capture and Yanomamo warfare: Retrospect and new evidence. *Journal of Anthropological Research* 40:183–201.

1984b   A cultural materialist theory of band and village warfare: The Yanomamo test. In *Warfare, culture, and environment*, edited by R. B. Ferguson, 111–40. Orlando: Academic Press.

Harrison, S.
1989   The symbolic construction of aggression and war in a Sepik River society. *Human Ecology* 24 (3): 413–26.

Harvey, D.
1996   *Justice, nature, and the geography of difference*. Cambridge: Blackwell.

Hass, J., ed.
1990   *The anthropology of war*. Cambridge: Cambridge University Press.

Hendrickson, D., J. Armon, and R. Mearns
1996   Livestock raiding among the pastoral Turkana of Kenya: Redistribution, predation, and the links to famine. *IDS Bulletin* 27 (3): 17–30.

1998   The changing nature of conflict and famine vulnerability: The case of livestock raiding in Turkana District, Kenya. *Disasters* 22 (3): 185–99.

Herskovits, M.
1926   The cattle complex in East Africa. *American Anthropologist.* 28:230–72.

Hobbes, T.
1656   *Leviathan.* London: William Hope.

Holling, C. S.
1973   Resilience and stability of ecological systems. *Annual Review of Ecology and Systematics* 4:1–23.

Holtzman, J. D.
2000   *Nuer journeys, Nuer lives: Sudanese refugees in Minnesota.* Boston: Allyn and Bacon.

Hrdy, S. B.
1999   *Mother Nature: A history of mothers, infants, and natural selection.* New York: Pantheon.

Hutchinson, S. E.
1996   *Nuer dilemmas: Coping with money, war, and the state.* Berkeley: University of California Press.

Ingold, T.
1980   *Hunters, pastoralists, and ranchers.* Cambridge: Cambridge University Press.
1987   *The appropriation of nature: Essays on human ecology and social relations.* Iowa City: University of Iowa Press.

Irons, W.
1979   Natural selection, adaptation, and human social behavior. In *Evolutionary biology and human social behavior: An anthropological perspective,* edited by N. Chagnon and W. Irons, 4–39. North Scituate, MA: Duxbury Press.

Irons, W., and N. Dyson-Hudson
1972   *Perspectives on nomadism.* Leiden: Brill.

Jacobs, A. H.
1979   Maasai inter-tribal relations: Belligerent herdsmen or peaceable pastoralists? In *Warfare among East African herders,* edited by K. Fukui and D. Turton, 33–52. Osaka: Senri Ethnological Series 3, National Museum of Ethnology.

Johnson, B. R., Jr.
1990   Nomadic networks and pastoral strategies: Surviving and exploiting local instability in South Turkana. Ph.D. dissertation, State University of New York.

Johnson, D.
1969   *The nature of nomadism: A comparative study of pastoral migrations in southwestern Asia and northern Africa.* Chicago: Department of Geography Research Paper 118, University of Chicago.

Keegan, J.
1997   War ça change. *Foreign Affairs* 76 (3): 113–16.

Keeley, L. H.
   1996   *War before civilization: The myth of the peaceful savage.* New York:
          Oxford University Press.
Kelly, R. C.
   2000   *Warless societies and the origin of war.* Ann Arbor: University of
          Michigan Press.
Kelly, R. L.
   1995   *The foraging spectrum: Diversity in hunter-gatherer lifeways.* Wash-
          ington, DC: Smithsonian Institution Press.
Kemp, W.
   1971   The flow of energy in a hunting society. *Scientific American*
          224:104–15.
Kenya Census
   1999   Kenya Census of Population and Housing: Tabulations. Republic
          of Kenya, Bureau of Statistics.
Khazanov, A. M.
   1994   *Nomads and the outside world.* Madison: University of Wisconsin
          Press.
Knauft, B. M.
   1990   Melanesian warfare: A theoretical history. *Oceania* 60:250–311.
Koch, K. F.
   1974   *War and peace in Jalemo: The management of conflict in highland New
          Guinea.* Cambridge: Harvard University Press.
Kottak, C. P.
   1999   The new ecological anthropology. *American Anthropologist* 101 (1):
          23–35.
Lamphear, J.
   1976   *The traditional history of the Jie of Uganda.* Oxford: Oxford Univer-
          sity Press.
   1988   The people of the Grey Bull: The origin and expansion of the
          Turkana. *Journal of African History* 29: 27–39.
   1992   *The scattering time: Turkana responses to colonial rule.* Oxford:
          Clarendon.
Lamprey, H.
   1983   Pastoralism yesterday and today: The overgrazing problem. In
          Ecosystems of the World, vol. 13, *Tropical Savannas,* edited by F.
          Bourliere, 643–66. Amsterdam: Elsevier Scientific.
Lattimore, O.
   1940   *Inner Asian frontiers of China.* Research Series 21. New York: Amer-
          ican Geographical Society.
Lee, R.
   1979   *The !Kung San: Men, women, and work in a foraging society.* Cam-
          bridge: Cambridge University Press.
Legge, K.
   1989   Changing responses to drought among the WodaaBe of Niger. In
          Bad year economics: Cultural responses to risk and uncertainty, edited

by P. Halstead and J. O'Shea, 81–86. Cambridge: Cambridge University Press.

Leslie, P. W., and B. C. Campbell
1995 Contrasting fertility patterns among nomadic and sedentary males in Turkana, Kenya. *American Journal of Physical Anthropology* Suppl. 20:135.

Leslie, P. W., K. Campbell, C. Campbell, C. Kigondu, and L. Kirumbi
1999 Fecundity and fertility. In *Turkana herders of the dry savanna: Ecology and biobehavioral response of nomads to an uncertain environment,* edited by M. A. Little and P. W. Leslie, 249–78. Oxford: Oxford University Press.

Leslie, P. W., K. L. Campbell, and M. A. Little
1993 Pregnancy loss in nomadic and settled women in Turkana, Kenya: A prospective study. *Human Biology* 65 (2): 237–54.

Leslie, P. W., and P. H. Fry
1989 Extreme seasonality of births among nomadic Turkana pastoralists. *American Journal of Physical Anthropology* 79:103–15.

Leslie, P. W., M. A. Little, R. Dyson-Hudson, and N. Dyson-Hudson
1999 Synthesis and lessons. In *Turkana herders of the dry savanna: Ecology and biobehavioral response of nomads to an uncertain environment,* ed. M. A. Little and P. W. Leslie, 355–73. Oxford: Oxford University Press.

Lindeman, R.
1942 The trophic dynamic aspect of ecology. *Ecology* 23:399–418.

Little, M. A., N. Dyson-Hudson, R Dyson-Hudson, J. E. Ellis, K. A. Galvin, P. W. Leslie, and D. M. Swift
1990 Ecosystem approaches in human biology: Their history and a case study of the South Turkana Ecosystem Project. In *The ecosystem approach in anthropology: From concept to practice,* edited by E. Moran, 389–434. Ann Arbor: University of Michigan Press.

Little, M. A., K. Galvin, and P. W. Leslie
1988 Health and energy requirements of nomadic Turkana pastoralists. In *Coping with uncertainty in food supply,* edited by I. de Garine and G. A. Harrison, 288–315. Oxford: Oxford University Press.

Little, M. A., K. Galvin, and M. Mugambi
1983 Cross-sectional growth of nomadic Turkana pastoralists. *Human Biology* 55 (4): 811–30.

Little, M. A., and S. J. Gray
1990 Growth of young nomadic and settled Turkana children. *Medical Anthropology Quarterly* 4 (3): 296–314.

Little, M. A., S. J. Gray, and P. W. Leslie
1993 Growth of nomadic and settled Turkana infants of northwest Kenya. *American Journal of Physical Anthropology* 92 (3): 273–89.

Little, M. A., and B. R. Johnson
1985 Weather conditions in South Turkana, Kenya. In *South Turkana nomadism: Coping with an unpredictably varying environment,* edited

by R. Dyson-Hudson and J. T. McCabe, 298–314. Ethnography Series FL17–001. New Haven: HRAFlex Books.

Little, M. A., and P. W. Leslie, eds.
1999 *Turkana herders of the dry savanna.* Oxford: Oxford University Press.

Little, P., K. Smith, B. Cellarius, L. Coppock, and C. Barrett
2001 Avoiding disaster: Diversification and risk management among East African herders. *Development and Change* 32 (3): 401–34.

Lizot, J.
1985 *Tales of the Yanomami: Daily life in the Venezuelan forest.* Cambridge: Cambridge University Press.
1994 Words in the night: The ceremonial dialogue—one expression of peaceful relationships among the Yanomami. In *The anthropology of peace and nonviolence,* edited by L. E. Sponsel and T. Gregor, 213–40. Boulder: Lynne Rienner.

Mace, R.
1990 Pastoralist herd compositions in unpredictable environments: A comparison of model predictions and data from cattle-keeping groups. *Agricultural Systems* 33:1–11.

Mace, R., and A. Houston
1989 Pastoralist strategies for survival in unpredictable environments: A model of herd composition that maximizes household variability. *Agricultural Systems* 31:185–204.

Markakis, J.
1994 Ethnic conflict and the state in the Horn of Africa. In *Ethnicity and conflict in the Horn of Africa,* edited by K. Fukui and J. Markakis, 217–37. London: James Currey.

McCabe, J. T.
1984 Food and the Turkana in Kenya. *Cultural Survival* 8 (1): 48–50.
1985 Livestock management among the Turkana: A social and ecological analysis of herding in an East African population. Ph.D. dissertation, State University of New York.
1987 Variation in livestock production and its impact on social organization. *Research in Economic Anthropology* 8:277–93.
1990 Success and failure: The breakdown of traditional drought coping institutions among the pastoralist Turkana of Kenya. *Journal of Asian and African Studies* 25 (3–4): 481–95.
1994 Mobility and land use among African pastoralists: Old conceptual problems and new interpretations. In *African pastoral systems: An integrated approach,* edited by E. Fratkin, K. A. Galvin, and E. A. Roth, 69–90. Boulder: Lynne Rienner.
2000 Patterns and processes of group movement in human nomadic populations: A case study of the Turkana of northwestern Kenya. In *On the move: How and why animals travel in groups,* edited by S. Boinski and P. Garber, 649–77. Chicago: University of Chicago Press.

2003 Sustainability and livelihood diversification among the Maasai of northern Tanzania. Manuscript. Submitted to *Human Organization* 62 (2): 100–111.

McCabe, J. T., C. Schofield, and S. L. Perkin
1992 Can conservation and development be coupled among pastoral people? The Maasai of the Ngorongoro conservation area. *Human Organization* 51 (4): 353–66.

McIntosh, R. P.
1985 *The background of ecology: Concept and theory.* Cambridge: Cambridge University Press.

Meggitt, M. J.
1977 *Blood is their argument: Warfare among the Mae-Enga tribesmen of the New Guinea highlands.* Palo Alto, CA: Mayfield.

Mirzeler, M., and C. Young
2000 Pastoral politics in the northeast periphery in Uganda: AK-47 as a change agent. *Journal of Modern African Studies* 38 (3): 407–29.

Montejo, V.
1987 *Testimony: Death of a Guatemalan village.* Willimantic, CT: Curbstone Press.

Moran, E., ed.
1984 *The ecosystem concept in anthropology.* Washington, DC: AAAS.
1990 Ecosystem ecology in biology and anthropology: A critical assessment. In *The ecosystem approach in anthropology: From concept to practice,* edited by E. Moran, 3–40. Ann Arbor: University of Michigan Press.

Morren, G.
1987 *The Miyanmin: Human ecology of a Papua New Guinea society.* Ann Arbor: UMI Research Press.

Mugambi, M.
1989 Responses of an African dwarf shrub, *Indigofera spinosa,* to competition, water stress, and defoliation. Ph.D. dissertation, Colorado State University.

Muller, H.
1989 *Changing generations; Dynamics of generation sets in southeastern Sudan (Toposa) and northwestern Kenya (Turkana).* Sarbrucken and Fort Lauderdale: Verlag Breitenbach.

Murray, M. G.
1995 Specific nutrient requirements and migration of wildebeest. In *Serengeti II: Dynamics, management, and conservation of an ecosystem,* edited by A. R. E. Sinclair and P. Arcese, 231–56. Chicago: University of Chicago Press.

Naimir-Fuller, M.
1994 Natural resource management at the local-level. In *Pastoral natural resource management and policy: Proceedings of the subregional workshop,* December 1993, Arusha. New York: UNESCO.

Netting, R. McC.
    1968   Introduction, *Hill farmers of Nigeria: A cultural ecology of the Kofyar of the Jos Plateau*. Seattle: University of Washington Press.
Nordstrom, C.
    1995   War on the front lines. In *Fieldwork under fire: Contemporary studies of violence and survival*, edited by C. Nordstrom and A. Robben, 128–53. Berkeley: University of California Press.
    1997   *A different kind of war story*. Philadelphia: University of Pennsylvania Press.
Nordstrom, C., and A. Robben, eds.
    1995   *Fieldwork under fire: Contemporary studies of violence and survival*. Berkeley: University of California Press.
Noy-Mier, I.
    1973   Desert ecosystems: Environment and producers. *Annual Review of Ecology and Systematics* 4:25–51.
Odum, E.
    1953   *Fundamentals of ecology*. Philadelphia: Saunders.
    1971   *Fundamentals of ecology*. 3d ed. Philadelphia: Saunders.
Orlove, B.
    1980   Ecological anthropology. *Annual Review of Anthropology* 9:235–73.
Otterbein, K. F.
    1997   The origins of war. *Critical Review* 11:251–77.
    1999   A history of research on warfare in anthropology. *American Anthropologist* 101 (4): 794–805.
Patten, R.
    1992   Pattern and process in an arid tropical ecosystem: The landscape and system properties of Ngisonyoka Turkana, north Kenya. Ph.D. dissertation, Colorado State University.
Peet, R., and M. Watts
    1996   *Liberation ecologies: Environment, development, social movements*. London and New York: Routledge.
Pennycuik, C. J.
    1975   On the running of the gnu (*Connochaetes taurinus*) and other animals. *Journal of Experimental Biology* 63:775–800.
Pike, I. L.
    1995   Pregnancy outcomes among nomadic Turkana herders of northwestern Kenya. *American Journal of Physical Anthropology* Suppl. 20:172.
Pimm, S. I.
    1991   *The balance of nature? Ecological issues in the conservation of species and communities*. Chicago: University of Chicago Press.
Piot, C.
    1999   *Remotely global: Village modernity in West Africa*. Chicago: University of Chicago Press.

Podolefsky, A.
   1984 Contempory warfare in the New Guinea highlands. *Ethnology*
         23:73–87.
Pospisil, L.
   1994 I am very sorry I cannot kill you anymore: War and peace among
         the Kapauku. In *Studying war: Anthropological perspectives,* edited
         by S. P. Reyna and R. E. Downs, 113–26. Amsterdam: Gordon and
         Breach.
Pratt, D. J., and M. D. Gwynne
   1977 *Rangeland management and ecology in East Africa.* London: Hodder
         and Stoughton.
Prince, S. D., and C. J. Tucker
   1986 Satellite remote sensing of rangelands in Botswana II. NOAA
         AVHRR and herbaceous vegetation. *International Journal of Remote
         Sensing* 7:1555–70.
Rappaport, R.
   1968 *Pigs for the ancestors: Ritual in the ecology of a New Guinea people.*
         New Haven: Yale University Press.
   1971 The flow of energy in an agricultural society. *Scientific American*
         225:117–32.
   1984 *Pigs for the ancestors: Ritual in the ecology of a New Guinea people.* 2d
         enlarged ed. New Haven: Yale University Press.
   1990 Ecosystems, populations, and people. In *The ecosystem approach in
         anthropology: From concept to practice,* ed. E. F. Moran, 41–72. Ann
         Arbor: University of Michigan Press.
Rayne, H.
   1923 *The ivory raiders.* London: W. Heinemann.
Reid, R. S.
   1992 Livestock-mediated tree regeneration: Impacts of pastoralists on
         woodlands in dry tropical Africa. Ph.D. dissertation, Colorado
         State University.
Reyna, S. P., and R. E. Downs, eds.
   1994 *Studying war: Anthropological perspectives.* Langhorne, PA: Gordon
         and Breach.
Richardson, F.
   1992 Challenges in animal production in southern Africa. Inaugural
         lecture at the Department of Animal Sciences, University of
         Bophutatswana.
Roscoe, P. B.
   1996 War and society in Sepik New Guinea. *Journal of the Royal Anthro-
         pological Institute* 2 (4): 645–66.
Roth, E.
   1996 Traditional pastoral strategies in a modern world: An example
         from northern Kenya. *Human Organization* 55:219–24.

Rubinstein, R. A., and M. LeCron Foster
    1988   *The social dynamics of peace and conflict: Culture in international secu-*
          rity. Boulder: Westview.
Salzman, P. C.
    1969   Multi-resource nomadism in Iranian Baluchistan. Presented at the
          Annual Meetings of the American Anthropological Association,
          New Orleans.
    1971   Movement and resource extraction among pastoral nomads: The
          case of the Shah Nawazi Baluch. *Anthropological Quarterly*
          44:185–97.
Sandford, S.
    1983   *Management of pastoral development in the Third World.* Chichester:
          John Wiley and Sons.
Schneider, Harold
    1957   *The Pakot (Suk) of Kenya with special reference to the role of livestock in*
          the subsistence economy. Ann Arbor: University of Michigan Press.
Schroeder, R. A.
    1993   Shady practice: Gender and the political ecology of resource stabi-
          lization in Gambian garden/orchards. *Economic Geography* 69 (4):
          349–65.
    1996   "Re-claiming" land in the Gambia: Gendered property rights and
          environmental intervention. *Annals of the Association of American*
          Geographers 87 (3): 487–508.
Scoones, I.
    1993   Exploiting heterogeneity: Habitat use by cattle in the communal
          areas of Zimbabwe. *Journal of Arid Environments* 29:221–37.
    1994   *Comments on livestock, land use, and agricultural intensification in*
          Sub-saharan Africa, by D. Bourne and W. Wint. Pastoral Develop-
          ment Network Paper, No. 37b. London: Overseas Development
          Institute.
    1996   Introduction. In *Living with uncertainty: New directions in pastoral*
          development in Africa, edited by I. Scoones, 1–36. London: Interme-
          diate Technology Publications.
    1999   New ecology and the social sciences: What prospects for a fruitful
          engagement? *Annual Review of Anthropology* 28:479–507.
Shankman, P.
    1991   Culture contact, cultural ecology, and Dani warfare. *Man* 26:299–321.
Shell-Duncan, B.
    1994   Child fostering among nomadic Turkana pastoralists: Demo-
          graphic and health consequences. In *African Pastoralist Systems: An*
          Integrated Approach, edited by E. Fratkin, K. A. Galvin, and E. A.
          Roth, 147–64. Boulder: Lynne Rienner.
    1995   Impact of seasonal variation in food availability and disease stress
          on the health status of nomadic Turkana children: A longitudinal

analysis of morbidity, immunity, and nutritional status. *American Journal of Human Biology* 7:339–55.

Shelley, K.

1985   Medicines for misfortune: Diagnosis and health care among southern Turkana pastoralists of Kenya. Ph.D. dissertation, University of North Carolina.

Sheridan, T.

1988   Where the dove calls: The political ecology of a peasant corporate community. Tucson: University of Arizona Press.

Sillitoe, P.

1977   Land shortage and war in New Guinea. *Ethnology* 16:71–81.

1978   Big men and war in New Guinea. *Man*, n.s., 13:252–71.

Simons, A.

1999   War: Back to the future. *Annual Review of Anthropology* 28:73–108.

Sinclair, A. R. E., and J. M. Fryxell

1985   The Sahel of Africa: Ecology of a disaster. *Canadian Journal of Zoology* 63:987–94.

Sluka, J. A., ed.

2000   *Death squad: The anthropology of state terror.* Philadelphia: University of Pennsylvania Press.

Smith, E. A.

1991   *Inujjuamuit foraging strategies.* New York: Aldine de Gruyter.

Smith, E. A., and B. Winterhalder

1992   *Evolutionary ecology and human behavior.* New York: Aldine de Gruyter.

Soper, R. C.

1985   *A socio-cultural profile of Turkana District.* Nairobi: Institute of African Studies, University of Nairobi.

Sponsel, L. E.

1998   Yanomami: An arena of conflict and aggression in the Amazon. *Aggressive Behavior* 24 (2): 97–122.

2000   Response to Otterbein. *American Anthropologist* 102 (4): 837–40.

2002   On reflections on *Darkness in El Dorado. Current Anthropology* 43 (1): 149–52.

Steward, J.

1955   The concept and method of cultural ecology. In *The theory of culture change,* 30–42. Urbana: University of Illinois Press.

Stoddart, L. A., A. D. Smith, and T. W. Box

1975   *Range Management.* New York: McGraw-Hill.

Stonich, S.

1993   Linking development, population, and the environment: Perspectives and methods. In *I am destroying the land! The political ecology of poverty and environmental destruction in Honduras,* edited by S. Stonich, 17–28. Boulder: Westview.

Strathern, A.
   1992   Let the bow go down. In *War in the tribal zone: Expanding states and indigenous warfare*, edited by F. R. Brain and N. L. Whitehead, 229–50. Santa Fe: School of American Research Press.

Tansley, A. G.
   1935   The use and abuse of vegetational concepts and terms. *Ecology* 16:284–307.

Thomas, R. B.
   1976   Energy flow at high altitude. In *Man in the Andes: A multidisciplinary study of high altitude Quechua*, edited by P. T. Baker and M. Little. Stroudsburg, PA: Dowden, Huchinson.

Tierney, P.
   2000   *Darkness in El Dorado: How scientists and journalists devastated the Amazon*. New York: W. W. Norton.

Tornay, S.
   1979   Armed conflicts in the Lower Omo Valley, 1970–1976: An analysis from within Nyangatom society. In *Warfare among East African herders*, edited by K. Fukui and D. Turton, 97–117. Osaka: Senri Ethnological Series 3, National Museum of Ethnology.

Tucker, C. J., J. R. Townsend, and T. E. Goff
   1985   African land-cover classification using satellite data. *Science* 227:369–75.

Tucker, C. J., B. VanPraet, and A. Gaston
   1983   Satellite remote sensing of total dry matter production in the Senegalese Sahel. *Remote Sensing Environment* 13:461–74.

Turner, B. L.
   1997   Spirals, bridges and tunnels: Engaging human-environment perspectives in geography? *Ecumene* 4 (2): 196–217.

Turton, D.
   1979   War, peace, and Mursi identity. In *Warfare among East African herders*, edited by K. Fukui and D. Turton, 179–210. Osaka: Senri Ethnological Series 3, National Museum of Ethnology.

   1994   Mursi political identity and warfare: The survival of an idea. In *Ethnicity and conflict in the Horn of Africa*, edited by K. Fukui and J. Markakis, 15–31. London: James Currey.

Turton, D., ed.
   1997   *War and ethnicity: Global connections and local violence*. Rochester: University of Rochester Press.

Tuzin, D.
   1997   *The cassowary's revenge: The life and death of masculinity in a New Guinea society*. Princeton: Princeton University Press.

Vayda, A. P.
   1971   Phases of war and peace among the Marings of New Guinea. *Oceania*. 42:1–24.
   1976   *War in ecological perspective*. New York: Plenum.
   1983   Progressive contextualization: Methods for research in human ecology. *Human Ecology* 11:265–81.

1989 Explaining why the Marings fought. *Journal of Anthropological Research* 45:159–77.

1995 Failures of explanation in Darwinian ecological anthropology—Parts I and II. *Philosophy of the Social Sciences* 25 (2): 219–49; 25 (3): 360–75.

Vayda, A. P., and B. J. McCay

1975 New directions in ecology and ecological anthropology. In *Annual review of anthropology.* edited by B. J. Siegal, A. R. Beals, and S. A. Tyler. Palo Alto: Annual Reviews Inc.

Vayda, A. P., and R. Rappaport

1968 Ecology: Cultural and non-cultural. In *Introduction to cultural anthropology: Essays in the scope and methods of the science of man,* edited by J. A. Clifton, 476–98. Boston: Houghton Mifflin.

Vayda, A. P., and B. B. Walters

1999 Against political ecology. *Human Ecology* 27 (1): 167–79.

Walker, P. L.

2001 A bioarchaeological perspective on the history of violence. *Annual Review of Anthropology* 30:573–96.

Westoby, M., B. Walker, and I. Noy-Meir

1989 Opportunistic management of rangelands not at equilibrium. *Journal of Range Management* 42:266–74.

Whitehead, N. L.

2001 A history of research on warfare in anthropology. Reply to Keith Otterbein. *American Anthropologist* 102 (4): 834–37.

Wienpahl, J.

1984a The role of women and small stock among the Turkana. Ph.D. dissertation, University of Arizona.

1984b Women's roles in livestock production among the Turkana of Kenya. *Research in Economic Anthropology* 6:193–215.

Wiens, J.

1977 On competition and variable environments. *American Scientist* 65:590–97.

1984 On understanding a non-equilibrium world: Myth and reality in community patterns and processes. In *Ecological communities: Conceptual issues and the evidence,* edited by D. R. Strong, D. Simberloff, L. G. Abele, and A. B. Thisstle, 439–57. Princeton: Princeton University Press.

Wilmsen, E.

1989 Land filled with flies: A political ecology of the Kalahari. Chicago: University of Chicago Press.

Wilson, E. O.

1978 *On human nature.* Cambridge: Harvard University Press.

Wolf, E.

1972 Ownership and political ecology. *Anthropological Quarterly* 45 (3): 201–5.

Worster, D.

1990 The ecology of order and chaos. *Environmental History Review* 14 (1–2): 1–18.

Wrangham, R.
    1999   Evolution of coalitionary killing. *Yearbook of Physical Anthropology*
           1–29.
Zimmerer, K. S.
    1994   Human geography and the "new ecology": The prospect and
           promise of integration. *Annals of the Association of American Geog-*
           raphers 84 (1): 108–25.

Index